Experiencing Climate Change in Bangladesh

Experiencing Climate Change in Bangladesh

Vulnerability and Adaptation in Coastal Regions

SALIM MOMTAZ
School of Environmental and Life Sciences,
University of Newcastle, Ourimbah, NSW, Australia

MASUD IQBAL MD SHAMEEM
School of Environmental and Life Sciences,
University of Newcastle, Ourimbah, NSW, Australia

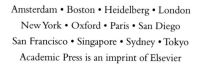

Amsterdam • Boston • Heidelberg • London
New York • Oxford • Paris • San Diego
San Francisco • Singapore • Sydney • Tokyo
Academic Press is an imprint of Elsevier

ELSEVIER

Academic Press is an imprint of Elsevier
125 London Wall, London EC2Y 5AS, UK
525 B Street, Suite 1800, San Diego, CA 92101-4495, USA
225 Wyman Street, Waltham, MA 02451, USA
The Boulevard, Langford Lane, Kidlington, Oxford OX5 1GB, UK

Notices
Knowledge and best practice in this field are constantly changing. As new research and
experience broaden our understanding, changes in research methods, professional practices,
or medical treatment may become necessary.

Practitioners and researchers must always rely on their own experience and knowledge in
evaluating and using any information, methods, compounds, or experiments described
herein. In using such information or methods they should be mindful of their own safety
and the safety of others, including parties for whom they have a professional responsibility.

To the fullest extent of the law, neither the Publisher nor the authors, contributors, or
editors, assume any liability for any injury and/or damage to persons or property as a
matter of products liability, negligence or otherwise, or from any use or operation of any
methods, products, instructions, or ideas contained in the material herein.

ISBN: 978-0-12-803404-0

British Library Cataloguing in Publication Data
A catalogue record for this book is available from the British Library

Library of Congress Cataloging-in-Publication Data
A catalog record for this book is available from the Library of Congress

For Information on all Academic Press publications
visit our website at http://store.elsevier.com/

Working together
to grow libraries in
developing countries

www.elsevier.com • www.bookaid.org

PREFACE

The natural resource–dependent societies of the coastal areas in Bangladesh have long been dealing with their vulnerabilities to extreme weather events. Marked changes in the coastal hydro-climatic environment are exacerbating the existing situation, with serious impacts on the environment, food production systems, and freshwater resources. As a result, coastal communities face the challenge of managing immediate livelihood threats and maintaining livelihood security in the long term. This book sets out to explore the processes by which rural households in coastal areas of Bangladesh adapt their livelihoods to climate variability and change within the context of a vulnerable setting. This book would be a valuable reference material for a wide range of reader including students, academics, professionals, and practitioners working in the field of climate change adaptation. Not only does it provide empirical information on climate change adaptation, but it also develops a theoretical conceptual framework to make meaningful interpretation of field data. We hope that this book will improve our understanding of how individuals or households adapt their livelihoods in response to climate variability, change, and extreme events.

ACKNOWLEDGMENTS

We are grateful to the Department of Education, Employment, and Workplace Relations of the Australian Government for providing an Endeavour Postgraduate Fellowship to carry out this research. We would like to thank the agency managing the Endeavour Fellowship, on behalf of the Australian Government, for their invaluable support. We would also like to thank the support staff at the School of Environmental and Life Sciences, University of Newcastle, for providing us with a supportive environment to conduct this research. Our special thanks go to the numerous individuals, villagers, officials, academics, and researchers who participated in our field investigations and provided valuable information.

CONTENTS

Preface   v
List of Figures   xi
List of Tables   xiii
About the Authors   xv
List of Acronyms   xvii

1. **Introduction**   **1**
   1.1 Problem Statement   1
   1.2 Investigative Questions   3
   1.3 Layout of the Book   4
   References   5

2. **Adaptation in Climate Change Discourse: A Conceptual Framework** **7**
   2.1 Introduction   8
   2.2 Adaptation in Theory   9
   2.3 Key Concepts in Climate Change Adaptation Studies   13
   2.4 Process of Adaptation to Climate Change   16
   2.5 Adaptation to Climate Change in Livelihood Framework   18
   References   25

3. **Study Design and Data Sources**   **29**
   3.1 Introduction   29
   3.2 Selection of Field Research Methods   30
   3.3 Study Design   31
   3.4 Case Study   34
   3.5 Field Data Collection   36
   References   39

4. **The Research Setting**   **41**
   4.1 Introduction   42
   4.2 Coastal Bangladesh   43
   4.3 Description of the Study Area   53
   4.4 Conclusion   65
   References   66

5. Household Assets and Capabilities   **69**

   5.1 Introduction   69
   5.2 Livelihood Capitals   70
   5.3 Discussion and Conclusion   82
   References   86

6. Local People's Perceptions of Climate Change   **87**

   6.1 Introduction   87
   6.2 Hydro-Climatic Variability and Extreme Climate Events   88
   6.3 Local Perceptions of Changes in Climate   93
   6.4 Comparison between Local Accounts of Climate Change and
        Meteorological Information   94
   6.5 Perceptions of Risks Concerning Climate Change   97
   6.6 Discussion and Conclusion   99
   References   101

7. Climate Disturbances and Change: Strategies for Adaptation   **103**

   7.1 Introduction   103
   7.2 Livelihood Diversification for Adaptation and Increasing Security   104
   7.3 Changing Livelihood Strategies for Adaptation to Climatic Hazards and
        Other Stressors   106
   7.4 Coping Strategies in Shrimp Aquaculture   110
   7.5 Adaptation to Salinity Intrusion in Rice Production   111
   7.6 Use of Climate Information   113
   7.7 Adaptation to Salinity Encroachment in Drinking Water Resources   114
   7.8 Improvement of Shelters: Households' Response to Tidal Flood   115
   7.9 Migration   116
   7.10 Discussion and Conclusion   117
   References   121

**8. Livelihood Adaptation to Climate Change: The Role of Policies
and Institutions**   **123**

   8.1 Introduction   123
   8.2 Institutional Interventions in Facilitating Adaptation to Climate Change   124
   8.3 Social Safety Nets: Public Responses to Cope with Livelihood Disturbances   125
   8.4 Role of NGOs in Promoting Livelihood Adaptation   129
   8.5 National Climate Policy and Livelihood Adaptation at the Local Level   132
   8.6 Adapting Development Plans and Sectoral Policies   133
   8.7 Discussion   134
   8.8 Conclusion   137
   References   138

9. Conclusion   **141**

9.1 Introduction   141

Reference   148

Glossary of Terms   *149*

Index   *151*

LIST OF FIGURES

Figure 2.1	Conceptualization of vulnerability according to the IPCC TAR	14
Figure 2.2	Process model of private proactive adaptation to climate change	17
Figure 2.3	Sustainable livelihoods framework	19
Figure 2.4	Conceptual framework for livelihood adaptation to climate variability and change	23
Figure 4.1	Map of the coastal region of Bangladesh	42
Figure 4.2	Types of coastal household by landholding size	46
Figure 4.3	Expansion of area under shrimp farming over 30 years	47
Figure 4.4	Map of the climatic subregions of Bangladesh	49
Figure 4.5	Waterlogging in coastal area	50
Figure 4.6	Map of historical cyclonic storm tracks	52
Figure 4.7	Maps showing location of Mongla Upazila in Bagerhat district and the study area	54
Figure 4.8	Distribution of population by age and gender	56
Figure 4.9	Distribution of households by farm type	59
Figure 4.10	Seasonal changes in salinity in the Mongla River	63
Figure 5.1	Household size and dependency ratio by income group	70
Figure 5.2	Self-assessed health status	72
Figure 5.3	Percentage of individuals received technical training	72
Figure 5.4	House types of the surveyed households	73
Figure 5.5	Lorenz curve of land distribution in the study area	75
Figure 5.6	Distribution of operated land by farm size in the study area	76
Figure 5.7	Percentage of households by flock size of poultry	77
Figure 5.8	Percentage of households by cattle (left) and goat (right) herd size	77
Figure 5.9	Income portfolios of surveyed households	78
Figure 5.10	Distributions of income and operated land	79
Figure 5.11	Lorenz curves of income distribution among surveyed households	79
Figure 5.12	Income portfolios across income strata, where Q (quintile) represents 20% of income distribution of all surveyed households	80
Figure 6.1	Box-and-whisker diagram of yearly temperatures in Mongla, 1989–2008. Whiskers represent maximum and minimum values, and boxes the interquartile range. The solid line within the box is the median	88
Figure 6.2	Box-and-whisker diagram of monthly total rainfall in Mongla, 1991–2008. Whiskers represent maximum and minimum values and boxes the interquartile range. The solid line within the box is the median and the continuous one, the mean	89
Figure 6.3	(a) Monthly distribution of major cyclonic storms over Bangladesh, 1960–2013. (b) Interdecadal trends in temporal distribution of major cyclone over Bangladesh, 1960–2009	90
Figure 6.4	Decadal frequency of cyclonic storms over the southwestern coast and all coastal regions of Bangladesh during 1960–2009	90

Figure 6.5   Decadal frequency of severe cyclones over the southwestern
coast and all coastal regions of Bangladesh during 1960–2009   91
Figure 6.6   Maximum wind speeds and corresponding surge heights of
the major cyclonic storms during 1960–2013   92
Figure 6.7   Decadal frequency of major cyclonic storms and associated
human casualties during 1960–2009   92
Figure 6.8   Annual departure of monsoon rainfall, 1991–2008 mean   96
Figure 6.9   Annual departure from winter rainfall, 1991–2008 mean   96
Figure 7.1   Income portfolios across income strata   104
Figure 7.2   Income portfolios across farming households   105
Figure 7.3   Changing patterns of livelihood activities over the last
10 years in Chila   107
Figure 7.4   A rainwater collection tank used by a household   115

LIST OF TABLES

Table 2.1	Bases for differentiating adaptation	12
Table 3.1	A list of variables representing the five livelihood capitals influencing the adaptive practices of rural households in the coastal area	33
Table 3.2	Categorization of household based on land holding size	36
Table 4.1	Basic demographic indicators of the national and coastal zone for the year 2011	44
Table 4.2	Major cyclones that hit the Bangladeshi coast, and death tolls	53
Table 4.3	Basic demographic indicators of Chila	55
Table 4.4	Public facilities in Chila	56
Table 4.5	Season calendar of livelihood activities in Chila	58
Table 4.6	Information on the damage in Chila caused by cyclone Aila	64
Table 4.7	Information on damages to fisheries sector and its estimated costs	64
Table 5.1	Rates of literacy in surveyed household members	71
Table 5.2	Sources of water used by survey population	74
Table 5.3	Agricultural and nonagricultural equipment owned by survey population	74
Table 5.4	Percentage of households by preferred source of credit	81
Table 6.1	Local perceptions of climate change	93
Table 6.2	Local perceptions of risk of climate change	98
Table 7.1	Adaptation measures implemented by shrimp farmer in Chila	110
Table 7.2	Percentage of households identifying the reasons for not using climate information	114
Table 8.1	Institutional interventions identified by local residents to be required for buffering impacts of climate change	125
Table 8.2	Overview of key social safety net programs (SSNs)	126
Table 8.3	An overview of NGO activities in Chila	130
Table 8.4	NGO-run programs participated in by survey households	132

ABOUT THE AUTHORS

Salim Momtaz is a senior lecturer at the University of Newcastle, Australia. He teaches in the area of sustainable resource management. He received a PhD from the University of London under a Commonwealth Scholarship. A geographer, an environmental scientist, and a social planner by training, Momtaz' research interests include environmental planning and governance, social adaptation to climate change, social impact assessment, and community engagement. His recently published books include *Sustainable Neighbourhoods in Australia: City of Sydney Urban Planning*, Springer (2015); *Brooklyn's Bushwick: Urban Renewal in New York, USA*, Springer (2014); *Sustainable Communities: A Framework for Planning*, Springer (2014); and *Evaluating Environmental and Social Impact Assessment in Developing Countries*, Elsevier (2013).

Masud Iqbal Md Shameem is a Deputy Director in the Department of Environment (DOE) in Bangladesh. In the DOE, he has nearly 20 years' experience in the fields of environmental impact assessment, monitoring, sustainability, and policy formulations. He also has experience in formulating and implementing projects related to hazardous waste management, water quality management, climate change, and biodiversity management. He has served on government committees dealing with various environmental issues. Shameem received a BSc in agriculture from Bangladesh Agricultural University and an MSc in environmental science from UNESCO-IHE Institute for Water Education in Netherlands under the Netherlands Fellowship Program. He has recently completed PhD research on livelihood adaptation to climate variability and change at the University of Newcastle in Australia under an Endeavour Postgraduate Fellowship of the Australian Government. Shameem has published a number of articles in international peer-reviewed journals.

LIST OF ACRONYMS

BCCRF	Bangladesh Climate Change Resilience Fund
BCCSAP	Bangladesh Climate Change Strategy and Action Plan
BCCTF	Bangladesh Climate Change Trust Fund
BDT	Bangladeshi Taka
BRRI	Bangladesh Rice Research Institute
CI	Corrugated Iron
CRED	Centre for Research on the Epidemiology of Disaster
DFID	Department for International Development
DRR	Disaster Risk Reduction
EMDAT	Emergency Event Database
HYV	High-Yielding Variety
IPCC	Intergovernmental Panel on Climate Change
IPCC TAR	Third Assessment Report of the Intergovernmental Panel on Climate Change
MOEF	Ministry of Environment and Forests
MPPACC	Model of Private Proactive Adaptation to Climate Change
NAPA	National Adaptation Plan of Action
NGO	Non-Governmental Organization
PL	Post Larvae
PMT	Protection Motivation Theory
SLA	Sustainable Livelihoods Approach
SP	Social Protection
SREX	Special Report on Managing the Risks of Extreme Events
SSN	Social Safety Network
UNFCCC	United Nations Framework Convention on Climate Change
VGD	Vulnerable Group Development
VGF	Vulnerable Group Feeding
WARPO	Water Resources Planning Organization

CHAPTER 1

Introduction

Contents

1.1 Problem Statement	1
1.2 Investigative Questions	3
1.3 Layout of the Book	4
References	5

1.1 PROBLEM STATEMENT

Bangladesh is widely regarded as one of the most vulnerable countries of the world to climate hazards (MOEF, 2009). The climatic vulnerability of the country can be attributed, in part, to the country's geographic position and its geomorphic conditions—both of which have made the country susceptible to river and rainwater flooding, tropical cyclones accompanied by storm surge, tornados, nor'westers, drought, and river bank and coastal erosion (Ali, 1996). The country is less than five metres above sea level, and about one-fourth of the country is flooded in an average year (MOEF, 2009). Bangladesh faced seven major floods between 1984 and 2012, and in 1998 nearly 68% of the country was inundated, affecting 30 million people, causing deaths to 1100 people, and incurring economic losses estimated to be US $2.8 billion (Bangladeshi Taka 135 billion) (MOEF, 2009).

The vulnerability arising from the hydro-climatic hazards is diverse in different parts of the country, and the community's ability to respond to these disturbances is highly variable. In Bangladesh, the coastal areas are considered as a zone of multiple vulnerabilities, exposed to climate extremes, viz. tropical cyclone, storm surge, and coastal flood. In the past two decades, of about 250,000 worldwide deaths associated with tropical cyclones, 60% occurred on the Bangladesh coast (ISDR, 2004). The coastal communities, especially those in the southwest region, are among the most vulnerable in the country, mostly due to their frequent exposure to tropical cyclones, with tidal surge and hydro-climatically driven changes in the environment and natural resource base on which they depend for their livelihoods and well-being. The southwest coastal region, therefore, presents a unique opportunity to explore how rural households and farmers who are faced with hydro-climatic risks manage their economic production systems and livelihoods.

Experiencing Climate Change in Bangladesh
http://dx.doi.org/10.1016/B978-0-12-803404-0.00001-6

Although social–ecological systems have evolved over the centuries through coping with, and adapting to, stresses and shocks, today increased pressure mainly associated with climate and environmental change poses a threat to coastal communities to manage their resources and livelihoods (UNEP, 2006; Parry et al., 2007). Adaptation to climate change, therefore, has emerged as a necessity for addressing the impacts of climate change on social well-being and the environment (Locatelli et al., 2008; Sonwa et al., 2010). In the rural livelihood–vulnerability context, individuals or households, when faced with climate risks, develop a suite of livelihood strategies over time, depending on their available assets, in pursuance of reducing their vulnerability and enhancing their livelihood resilience to climate shocks. Although successful livelihood adaptation contributes to an increasing well-being and livelihood security, maladaptation or failure in adaptation can result in persistent poverty (Tschakert and Dietrich, 2010). This realization of the importance of adaptation has promoted climate change adaptation research to the forefront of scientific inquiry and policy negotiations (Tschakert and Dietrich, 2010).

Policy response is an important institutional and governmental mechanism that strengthens adaptive capacity and expands adaptation options, as it may have a similar constraining effect (Smit et al., 2001; Urwin and Jordan, 2008). To support local responses to manage climate risks, Bangladesh has made significant progress in terms of policy development, institutional change, funding mechanisms, and implementation of a number of structural and nonstructural interventions. There are also an increasing number of community-based adaptation programs, mostly being carried out by non-governmental organizations (NGOs) on rural livelihood adaptation to hydro-climatic extreme events such as floods, cyclones, and storm surges (IDS, 2011). In these circumstances, it is important to build up the evidence of adaptations that are taking place on the ground and to identify the processes of effective adaptation that enhance livelihood resilience, promote legitimate institutional change, and support synergy with other goals of sustainable development (Robinson et al., 2006; Osbahr et al., 2008).

The vulnerability and low adaptive capacity of Bangladesh has attracted significant local and international interest in conducting research on climate change adaptation in the country. However, the predominant approach in adaptation research is either descriptive (for example, listing the elements of adaptive capacity that influence adaptation) or normative (wish lists of policy entry points), rather than analytical approaches that systematically examine the processes of adaptation (Berkhout et al., 2004; Berrang-Ford et al., 2011).

Drawing on the body of literature on climate change adaptation published in the journal *Climatic Change* between 1977 and 2010, Arnell (2010) proposes that future research on climate change adaptation should be focused on local contexts and address, and so forth, the process by which adaptations to climate change occur. Arnell further proposes that the model of adaptation should be constructed by taking account of local circumstances, encompassing geophysical characteristics, governance and management practices, and institutional context, all of which significantly affect the actual decision-making processes relating to adaptation.

The research agendum for future adaptation research proposed by Arnell (2010) runs parallel with the observation made in an adaptation-related chapter of the Fourth Assessment Report of the Intergovernmental Panel on Climate Change (IPCC AR4) (Adger et al., 2007, p. 737). The report concludes that there are significant knowledge gaps in understanding the processes by which adaptation is taking place and in identifying the areas for leverage and action by government.

1.2 INVESTIGATIVE QUESTIONS

To address the research problem, as stated above, this study aims to empirically explore the processes that facilitate adaptation to climate variability and change within the livelihood-vulnerability context of rural households in the coastal areas of Bangladesh. The study investigates the important livelihood assets and relevant institutional and policy contexts that provide rural households with the capacity to adapt their livelihood systems to climate variability and change. It explores the question of communities' perceptions of climate change, as a precondition for making a decision to adapt, and identifies actual adaptive responses by households (adaptations) in relation to these individuals' perceptions of climate change. In particular, this research intends to answer the following five specific research questions:

1. What are the key features of the livelihoods of the households that result in the ability of the households to cope with, and adapt to, climate variability and change?

2. What are the perceptions of the local communities about historic hydro-climatic variations, changes and extreme climate events, and their corresponding impacts on different livelihood assets, activities, and outcomes?

3. What are the climate-related risks, uncertainties, and opportunities that are perceived by local people as affecting their livelihoods?

4. What are the coping and adaptation strategies that people undertake to maintain resilient livelihoods in the face of climate variability and change?
5. In what ways do different institutions, agencies and policies influence livelihood adaptation to climate change disturbances in the coastal areas?

1.3 LAYOUT OF THE BOOK

Chapter 2 reviews the concepts associated with adaptation to climate change, and provides an overview of adaptation models based on a cognitive perspective to explain the process of adaptation at an individual level. It discusses the Sustainable Livelihoods Approach (SLA), and, by combining insights from the Socio-cognitive Model of Adaptation with SLA, a conceptual framework for livelihood adaptation to climate change is adopted and used to guide the research. This conceptual framework illustrates the key features of the livelihood adaptation process of rural households in response to changes in the climatic characteristics.

Chapter 3 presents the research design, data collection, and analysis. It explains the rationale for the mixed study design adopted for the study. The chapter also describes the procedures for selecting a case study to apply the research methodology to investigate the adaptation process in the coastal area facing the effects of climate-related events.

Chapter 4 begins by presenting an overview of the physical setting, socio-economic conditions, and hydro-climatic hazards of the coastal areas of Bangladesh that form the large-scale context in which households develop their livelihood strategies to adapt to climate hazards. It also presents background information about the relevant climate features and discusses the potential changes in the climate regime that has emerged from recent studies. The second part of this chapter serves as an introduction to the case study area, with information on its geographic location and socio-economic features, including its population dynamics, the livelihood activities of the households, public infrastructure, and services.

Chapter 5 examines households' access to the five main types of asset (human, social, natural, physical, and financial) as providing the potential capabilities of the households to integrate climate risk management in their livelihood strategies.

Chapter 6 discusses the local people's perceptions of the changing climate around them affecting their livelihoods. These perceptions largely influence adaptation decisions regarding the choice of a specific strategy or a set of strategies to enhance their existing livelihood security or reduce their vulnerability.

Chapter 7 explores whether, and to what extent, the perception of climate risk and the livelihood assets accessed by the households contribute to developing an adaptive livelihood practice to reduce vulnerability and enhance resilience to the impacts of climate variability and change. This chapter incorporates information and insights gained through Chapters 5 and 6, and illustrates how the adaptation process, triggered by climate risk perception, can, on one hand, motivate households to persue diversified livelihood strategies; on the other hand, the adaptation process can lead to incremental adjustments in the routine livelihood activities in order to withstand and recover from short-term weather-related shocks and changes in the long-term climatic conditions.

Chapter 8 examines policies, institutions, and practices, from the perspectives of vulnerability reduction and stimulating adaptation, in relation to climate variability, extreme events, and change. It looks at aspects of the social protection strategies and disaster reduction policies that enhance the capacity of the coastal communities to protect their livelihoods from climate hazards and to adapt to long-term changes in climate conditions. The chapter ends by looking at the government's policies and practices related to climate change adaptation and their implications in supporting livelihood adaptation in the coastal areas.

Chapter 9 summarizes the key findings that emerge from the preceding chapters. It revisits the questions and presents the answers to those questions. The chapter then re-examines the conceptual framework for livelihood adaptation to climate change and discusses its practical use to investigate the process of livelihood adaptation at a household level.

The final section of the book concludes the study by presenting the policy implications of the research with regard to facilitating adaptation to climate change in the coastal areas of Bangladesh.

REFERENCES

Ali, A., 1996. Vulnerability of Bangladesh to climate change and sea level rise through tropical cyclone and sea level rise. Water Air Soil Pollut. 92, 171–179.

Arnell, N.W., 2010. Adapting to climate change: an evolving research programme. Clim. Change 100, 107–111.

Adger, W.N., Agrawala, S., Mirza, M.M.Q., Conde, C., O'Brien, K., Pulhin, J., Pulwarty, R., Smit, B., Takahashi, K., 2007. Assessment of adaptation practices, options, constraints and capacity. In: Parry, M.L., Canziani, O.F., Palutikof, J.P., van der Linden, P.J., Hanson, C.E. (Eds.), IPCC Climate Change 2007: Impacts, Adaptation and Vulnerability. Contribution of Working Group II to the Fourth Assessment Report of the Intergovernmental Panel on Climate Change. Cambridge University Press, Cambridge, pp. 717–743.

Berkhout, F., Hertin, J., Gann, D., 2004. Learning to Adapt: Organizational Adaptation to Climate Change Impacts. Tyndall Centre Technical Report No. 11. Tyndall Centre for Climate Change Research, UK.

Berrang-Ford, L., Ford, D.J., Paterson, J., 2011. Are we adapting to climate change? Glob. Environ. Change 21, 25–33.

ISDR, 2004. Living with Risk: A Global Review of Disaster Reduction Initiatives. International Strategy for Disaster Reduction (ISDR).

IDS, 2011. Climate Finance in Bangladesh: Lessons for Development Cooperation and Climate Finance at National Level. Institute of Development Studies. Working Paper 5280.

Locatelli, B., Markku, K., Brockhaus, M., Colfer, C.J.P., Murdiyarso, D., Santoso, H., 2008. Facing an Uncertain Future. How Forests and People Can Adapt to Climate Change. Centre for International Forestry Research, Indonesia. p. 100.

MOEF, 2009. Bangladesh Climate Change Strategy and Action Plan 2009. Ministry of Environment and Forests, Bangladesh.

Osbahr, H., Twyman, C., Adger, W.N., Thomas, D.S.G., 2008. Effective livelihood adaptation to climate change disturbance: scale dimensions of practice in Mozambique. Geoforum 39, 1951–1964.

Parry, M.L., Canziani, O.F., Palutikof, J.P., van der Linden, P.J., Hanson, C.E., 2007. Contribution of Working Group II to the Fourth Assessment Report of the Intergovernmental Panel on Climate Change. Cambridge University Press, Cambridge, UK and New York, NY, USA.

Robinson, J., Bradley, M., Busby, P., Connor, D., Murrey, A., Sampson, B., Soper, W., 2006. Climate change and sustainable development: realizing the opportunity. Ambio 35 (1), 2–8.

Sonwa, D.J., Bele, M.Y., Somorin, O.A., Nkem, J., 2010. Central Africa is not only carbon stock: preliminary efforts to promote adaptation to climate change for forest and communities in Congo Basin. Nat. Faune 23 (1), 55–57.

Smit, B., Pilifosova, O., Burton, I., Challenger, B., Huq, S., Klein, R.J.T., Yohe, G., 2001. Adaptation to climate change in the context of sustainable development and equity. In: McCarthy, J.J., Canziani, O., Leary, N.A., Dokken, D.J., White, K.S. (Eds.), Climate Change 2001: Impacts, Adaptation and Vulnerability. Contribution of the Working Group II to the Third Assessment Report of the Intergovernmental Panel on Climate Change. Cambridge University Press, Cambridge, pp. 877–912.

Tschakert, P., Dietrich, K.A., 2010. Anticipatory learning for climate change adaptation and resilience. Ecol. Soc. 15 (2), 11.

UNEP, 2006. Marine and Coastal Ecosystems and Human Well-being: A Synthesis Report Based on the Findings of the Millennium Ecosystem Assessment. UNEP. p. 76.

Urwin, K., Jordan, A., 2008. Does public policy support or undermine climate change adaptation? exploring policy interplay across different scales of governance. Glob. Environ. Change 18, 180–191.

CHAPTER 2

Adaptation in Climate Change Discourse: A Conceptual Framework

Contents

2.1	Introduction	8
2.2	Adaptation in Theory	9
	2.2.1 Evolution of Approaches to Adaptation	9
	2.2.2 A Conceptual Framework of Adaptation to Climate Change	10
	2.2.2.1 Climate Stimuli	*10*
	2.2.2.2 System Definition	*10*
	2.2.2.3 Adaptive Responses	*11*
2.3	Key Concepts in Climate Change Adaptation Studies	13
	2.3.1 Vulnerability to Climate Change	13
	2.3.2 Resilience Framework in Climate Change Adaptation	14
	2.3.3 Maladaptation	15
2.4	Process of Adaptation to Climate Change	16
	2.4.1 Strength of Belief as Motivation to Climate Change Adaptation	16
	2.4.2 Socio-Cognitive Model of Adaptation to Climate Change	17
2.5	Adaptation to Climate Change in Livelihood Framework	18
	2.5.1 Sustainable Livelihoods Approach (SLA)	19
	2.5.1.1 The Role of Assets	*20*
	2.5.1.2 Vulnerability Context	*20*
	2.5.1.3 The Role of Institution and Structures	*20*
	2.5.1.4 Livelihood Strategies and Outcomes	*21*
	2.5.2 Adapting Livelihood Approaches to Climate Change	22
	2.5.3 A Conceptual Framework for Livelihood Adaptation to Climate Variability and Change	23
	2.5.3.1 Access to Livelihood Assets	*24*
	2.5.3.2 Climate Risk Perception	*24*
	2.5.3.3 Embedding Climate Risk Management into Livelihood Strategies	*24*
	2.5.3.4 Institutional Context	*24*
	2.5.3.5 Climate-Resilient Livelihood Outcomes	*25*
	2.5.3.6 Feedback Mechanisms	*25*
References		25

Experiencing Climate Change in Bangladesh
http://dx.doi.org/10.1016/B978-0-12-803404-0.00002-8

7

2.1 INTRODUCTION

There is substantial evidence that the climate is changing (Cubasch et al., 2013). Historical greenhouse gas (GHG) emissions have already "committed" the earth to some level of warming (Adger et al., 2007), and the global mean temperature will probably exceed 2 °C against 1900 level over the next decades, regardless of mitigation measures (Parry et al., 2009). Global warming beyond this threshold level (2 °C against 1900 level) is considered to be dangerous, in that this could interfere with the climate system and risk very large impacts on multicentury time scales (Parry et al., 2009; Smith et al., 2009). These changes are leading to environmental impacts such as global average sea level rise, changes in temperature and precipitation extremes, and changes in tropical cyclones. Many of these changes will lead to multiple socio-economic impacts such as altering today's yields, earning, health and physical safety and, ultimately, the paths and levels of future development (World Bank, 2010). Although climate change will affect everyone, it is expected to have a disproportionate effect on those who live in poverty in developing countries (POST, 2006).

In the effort to grapple with the challenge of global climate change, adaptation is unavoidable, as the most restricted measures to reduce the emission of greenhouse gases (GHGs) at this stage would not be sufficient to avoid the impacts of climate change (Berrang-Ford et al., 2011). Some suggest that the challenge of adaptation is not new, as societies have adapted to climate change over the course of human history. Empirical studies show that many individuals and communities within a society show clear signs of buffering their livelihoods in the face of disturbance and having the capacity to adjust their livelihood pathways to moderate the effects of climate change (Osbahr et al., 2008; Thomas, 2008).

The purpose of this chapter is to review the literature in the field of climate change and livelihoods, and to set out a conceptual framework to guide the research. In the next section, the concept of adaptation in climate change discourse is introduced, including a brief history of this concept. Other, related concepts explored to complement the adaptation framework include vulnerability, resilience, and maladaptation. This chapter provides an overview of the adaptation models built on the different theoretical perspectives that provide a foundation to understand the adaptation process taking place in socio-ecological systems under climate change risks. The chapter begins with a discussion of the concepts of the livelihood approach, which is then integrated with approaches to climate change adaptation in order to understand the conceptual and methodological linkage between adaptation and livelihood systems. Finally, a conceptual framework on the livelihood adaptation process is proposed, by combining insights from both adaptation livelihood frameworks, which is then used to structure the research underpinning this book and to answer the research questions presented in Chapter 1.

2.2 ADAPTATION IN THEORY

2.2.1 Evolution of Approaches to Adaptation

The term "adaptation" is receiving increasing attention from people who are concerned about climate change. Although the term has proliferated recently in the context of climate change, the nontechnical meaning of this word came into use in the English language in the early seventeenth century (Orlove, 2009). Yet, the word has acquired specific meaning in particular disciplines. To comprehend how adaptation is defined and understood, both in current discourses and in policies, it is helpful to first trace the inception of this term and to investigate the conceptual evaluation of adaptation.

According to dictionaries, "adapt" means to change something, to make it suitable for different conditions, and "adaptation" refers to the process of changing to suit different conditions. Adaptation appeared as a technical term first in evolutionary biology and was notably used by Darwin in his 1859 book *On the Origin of Species* (Orlove, 2009). According to Darwin's theory of evolution by natural selection, adaptation refers to the organic modification by which an organism or species becomes fitted to its environment. In biology, adaptation refers to: (1) physiological changes for adjusting to an immediate environment; (2) the process of becoming adapted, driven by genetic variations among individuals that enable an organism's survival and reproduction in a specific environment; and (3) development of a particular feature through evolution by natural selection for a specific function (Encyclopedia Britannica, 2011).

Influenced by evolutionary biology, the concept of adaptation to human systems came into use, either explicitly or implicitly in a number of social science fields, from welfare economics and anthropology to human geography and political ecology. The concept of adaptation in the field of climate change evolved concurrently with the increasing concern about climate variability and change. Although there are a number of definitions of adaptation in the climate change literature, adaptation refers usually to an adjustment in a system in response to climatic stimuli. Efforts have been made to develop common definitions and a coherent conceptual framework of adaptation through the Intergovernmental Panel on Climate Change (IPCC). The IPCC's Special Report on Managing the Risks of Extreme Events and Disasters to Advance Climate Change Adaptation (SREX) has defined adaptation as the process of adjustment in natural and human systems to reduce damage or to exploit beneficial opportunities in response to real or expected climate and associated effects (IPCC, 2012). The IPCC conceptualizes adaptation in the context of the adaptive capacity and vulnerability of ecological–socio-economic systems to climate change.

2.2.2 A Conceptual Framework of Adaptation to Climate Change

Drawing on Smit et al. (2000), the Third Assessment Report (TAR) of the IPCC identifies three major dimensions of adaptation that together offer a conceptual framework of adaptation (Smit et al., 2001).

2.2.2.1 Climate Stimuli
Berkhout et al. (2004) state that climate stimuli are those features of climate that have some influence on the behavior of a system. Other terms used to express climate stimuli include stresses, disturbances, events, hazards, and perturbations. Climate-related stimuli for which adaptation is undertaken include changes in average yearly weather conditions (e.g., temperature, precipitation), great variability within the range of normal climatic conditions (interannual variation), and changes in extreme events or catastrophic weather conditions such as floods, droughts, or storms (partly based on Smit et al., 2001).

2.2.2.2 System Definition
The SREX of the IPCC states that "adaptation occurs in natural and human systems" (IPCC, 2012, p. 5). The system of interest is also termed a "unit of analysis," "exposure unit," "activity of interest," or "sensitive system" (Smit et al., 2001). The characteristics of the system influence the occurrence and nature of the adaptations. The following are the important general properties of human–environment systems that are pertinent to adaptation (Smithers and Smit, 2009; Smit et al., 2000):

> **Sensitivity:** refers to the degree to which a system is modified or affected by disturbance (Adger, 2006). A stable system has the capacity to absorb disturbance so as to fluctuate little in response to climate change.

Resilience: the capability of a system to self-organize while undergoing change and still retain the same controls on function and structure. Berkes et al. (2003) note three characteristics of resilience: capacity to (1) respond to disturbance, (2) self-organize, and (3) learn and adapt.

Vulnerability: refers to the susceptibility of a system to suffer harm as a result of adverse effects of climate change. The key parameters of vulnerability are the exposure of a system to climate stimuli and adaptive capacity of the system to deal with stimuli.

Adaptive capacity: the potential or ability of a system to respond successfully to climate variability and change (Adger et al., 2007). Adaptive capacity amounts to the capacity of the human actors to manage the resilience of the social-ecological systems (Walker et al., 2004), and adaptation in this regard is the manifestation of the adaptive capacity.

Scale: the scale of a system of interest matters. For example, institutions and actors work across scale. Cash et al. (2006) define scale as "the spatial, temporal, quantitative, or analytical dimensions used to measure and study any phenomenon, and levels as the units of analysis that are located at different positions on a scale". Different components of a human–environmental system, such as temporal, jurisdictional, and institutional issues, can be considered at different levels of their scales. Osbahr et al. (2008) observed that complex cross-scale interactions are essential to deal with environmental problems, such as climate change, which is a global phenomenon but has local outcomes.

2.2.2.3 Adaptive Responses

Much of the scholarship in the field of adaptation research has identified the important attributes of adaptation so as to distinguish the different types of adaptation. Drawing on Smit et al. (2000), the IPCCTAR summarizes seven broad attributes that provide a base for characterizing and differentiating adaptations, as shown in Table 2.1.

Climate change adaptation is often differentiated between autonomous and planned adaptation. Autonomous or spontaneous adaptation takes place as a reactive response to climate stimuli (Smit et al., 2001). This type of adaptation does not constitute a conscious response to climatic stimuli but is triggered by ecological changes in natural systems and by market or welfare changes in human systems. Autonomous adaptation is considered to be undertaken by private sector and individual initiatives. On the other hand, planned adaptations are deliberately planned strategies that are often undertaken by the public sector. These can be anticipatory or reactive, but are

Table 2.1 Bases for differentiating adaptation (Smit et al., 2000)

General attribute	Examples of terms used	
Purposefulness	Autonomous	Planned
	Spontaneous	Purposeful
	Automatic	Intentional
	Natural	Policy
	Passive	Active
		Strategic
Timing	Anticipatory	Responsive
	Proactive	Reactive
	ex ante	*ex post*
Temporal scope	Short-term	Long-term
	Tactical	Strategic
	Instantaneous	Cumulative
	Contingency	
	Routine	
Special scope	Localized	Widespread
Function/effects	Retreat, accommodate, protect	
	Prevent, tolerate, spread, change, restore	
Form	Structural, legal, institutional, regulatory, financial, technological	
Performance	Cost-effectiveness, efficiency, implementability, equity	

based on an awareness that conditions have changed or are about to change and that action is required to maintain or to achieve a desired state. Therefore, autonomous and planned adaptations are often regarded as private and public adaptations, respectively. Understanding the autonomous adaptation is particularly important in designing the adaptation policy options, i.e., planned adaptation. The analysis of autonomous adaptation helps to identify major policies (e.g., market stimuli, technology discrimination), contributing to self-motivated adaptation in the society, which can be benchmarked to further advance the adaptation process by planned interventions so as to make the process effective and sustainable.

Depending upon the time of response actions, adaptation can be reactive or anticipatory. Anticipatory or proactive adaptation takes place before the manifestation of the impacts of climate change, whereas reactive adaptation occurs after the initial impact of climate change is observed. Adaptation can be distinguished according to its time frame, such as short-term (tactical) versus long-term (strategic) responses to climatic conditions. Tactical adaptation is often interpreted as a coping strategy, such as the selling of livestock during drought or flood. Strategic adaptation refers to the structural changes in a system that are applied over a long period of time, such as changes in land use, crop type and the use of insurance (Smit and Skinner, 2002). Further distinguishing features of adaptation are the scale at which they occur and the actor responsible for adaptive response. Adaptation involves a wide range of decision makers, both in the private (individuals, households, businesses, and corporates) and public spheres, at different hierarchical level. The IPCC TAR highlights the importance of distinguishing the various decision makers involved in the adaptation process because each stakeholder has a distinct capability to consider a distinct type of adaptive response.

2.3 KEY CONCEPTS IN CLIMATE CHANGE ADAPTATION STUDIES

2.3.1 Vulnerability to Climate Change

Adaptation is a deliberate process of change in response to real or perceived potential for damage or harm. This potentiality to be harmed is described by the term "vulnerability." As such, this term is a core concept of adaptation research; understanding vulnerability is therefore critical in the exploration of adaptive responses to climate-related hazards and the changing trends in human–environment systems.

In simple terms, vulnerability is used to describe the condition of susceptibility to be harmed. It is often conceptualized as a function of the character and magnitude of stressors to which a system is exposed, its sensitivity, and its capacity to deal with the effects of these stressors (IPCC, 2001). This definition suggests that vulnerability relates to the sources of stressors, which are external to a system, and to a system itself that must seek to cope with them. The relevant system may be human, such as an individual or a population; a business enterprise or an entire economy; or an ecological system including a single species or an entire ecosystem.

The term "vulnerability" is used by a host of research communities including livelihoods, food security, natural hazards, disaster risk management, public health, global environmental change, and climate change, yet with different conceptualizations and framings. This plurality of disciplines in the field of vulnerability to climate change has led to a diversity in definitions of the term, which is accompanied by a similar diversity of methodologies for assessing vulnerability (Hinkel, 2011). Within this terminological and methodological confusion associated with vulnerability, it is worth examining the interpretation of vulnerability to climate change developed within the assessment report of the Intergovernmental Panel on Climate

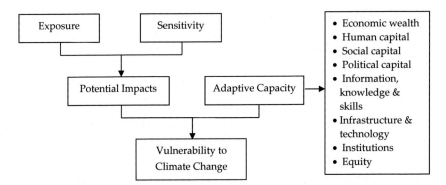

Figure 2.1 Conceptualization of vulnerability according to the IPCC TAR. *Source: Compiled from Ionescu et al. (2009).*

Change (IPCC), arguably the most authoritative source in the context of climate change. In the Third Assessment Report (TAR) of the IPCC, the vulnerability to climate change was described as follows: "the degree to which a system is susceptible to, and unable to cope with, adverse effects of climate change, including climate variability and extremes. Vulnerability is a function of the character, magnitude, and rate of climate change and variation to which a system is exposed, its sensitivity, and its adaptive capacity" (IPCC, 2001, p. 995). Based on this IPCC definition, the relationship among the primary determinants of vulnerability is illustrated in Figure 2.1.

Article 4.4 of the United Nations Framework Convention on Climate Change (UNFCCC) calls on developed countries to "assist the developing country parties that are particularly vulnerable to the adverse effects of climate change in meeting costs of adaptation" (United Nations, 1992, p. 14). To fulfil the obligation set out in this article, understanding vulnerability is crucial so as to identify adaptation needs and to inform policy development.

2.3.2 Resilience Framework in Climate Change Adaptation

The resilience concept is increasingly being used in the discourses on climate change adaptation, as it helps to understand the processes of change and the long-term trajectory of social–ecological systems (Nelson, 2011). Within the context of climate variability and change, the purpose of adaptation is often considered as performed to reduce vulnerability or to enhance resilience of the human–environment systems (Smit et al., 2001). The system's resilience, as it applies to social–ecological systems (SES), has three defining characteristics: (1) ability to absorb disturbance; (2) capability to self-organization; and (3) ability to learn and adapt to changes (Berkes et al., 2003). The magnitude, type, and complexity of changes in climate, society, and ecosystems are not always predictable, but change will occur. Resilience theory was developed to deal with these processes of change, focusing on the long-term sustainability within social–ecological systems (Adger et al., 2011).

The resilience framework is based on the theory of complex systems, recognizing that an ecological system is intricately linked with, and affected by, social systems. Both the social and ecological systems are connected through their interactive subsystems that are operating on different scales, ranging from households to village to nations, and from trees to patches to landscapes (Walker et al., 2004). Thus, resilience thinking does not imply a focus merely on ecosystems or societies but, rather, on coupled social–ecological system, underscoring the two-way interactions between human and natural system (Berkes, 2007). When this idea applies in the context of climate change, it implies that a society, in response to climate risks, may be able to take adaptation measures through simple technological means, but that it may not be able to meet the long-term adaptation objectives unless the overall system resilience is evaluated by taking full account of the ecological perspective, focusing on the long-term sustainability of the social–ecological systems.

2.3.3 Maladaptation

Although there has been much attention focused on promoting adaptation in reducing vulnerability stemmed from climate variability and change, there is also the possibility that adaptation actions can exacerbate the vulnerability to climate change–related hazards, or that they may fail to meet expected outcomes. This is termed "maladaptation." Barnett and O'Neill (2010, p. 211) defined maladaptation as follows: "action taken ostensibly to avoid or reduce vulnerability to climate change that impacts adversely on, or increases the vulnerability of other systems, sectors or social groups." In an investigation to adaptation responses to water stress in Melbourne, Australia, they identified five distinct types or pathways through which maladaptation to climate change arises. These are the following:
- Increased emission of greenhouse gases (e.g., increased use of energy-intensive air conditioners in response to heat waves)
- Disproportionately burdening the most vulnerable (e.g., costs of adaptation interventions impacts poor households disproportionately)
- High opportunity cost (e.g., an adaptation option that has higher economic, social and environmental cost than other alternative options)

- Reduced incentive to adapt
- Path dependency (e.g., inflexible institutions or large investment that reduces the possibility for future adaptation)

Tschakert and Dietrich (2010) mention the risk of maladaptive undertakings in the livelihood context of poor pastoralist communities in eastern Africa. They highlight that any unsuccessful attempt to manage climate risk by poor communities can trap them in chronic poverty.

2.4 PROCESS OF ADAPTATION TO CLIMATE CHANGE

Adaptation is a process of making and implementing decisions by an individual, organization, or government with regard to managing risks related to changes in climate, using the adaptive capacity available to those actors (Nelson, 2011). The Fourth Assessment Report of the IPCCC emphasizes the presence of an adaptive capacity for designing and implementing effective adaptation strategies (Adger et al., 2007). The factors related to general adaptive capacities include economic wealth, social networks, and equity, whereas in the operation context, the related factors include knowledge and skills, access to resources and technology, and institutional supports (Wesche and Armitage, 2010). Scholars argue that developing economies, with limited access to information and technology, poor infrastructure, low level of skills, weak institutions, and inequalities in power and access to resources, have little capacity to adapt and are thus vulnerable to climate change (Grothmann and Patt, 2005). However, based on empirical research, numerous authors note that adaptive capacity alone does not automatically trigger adaptation (Naess et al., 2005). Therefore, a central component of recent discussion has been focused on the question of what influences the use of adaptive capacity to produce adaptation in response to climate variability and change.

2.4.1 Strength of Belief as Motivation to Climate Change Adaptation

To understand the factors responsible for translating adaptive capacity into adaptation, Blennow and Persson (2009) state that economic, social, and political arrangements cannot alone explain the adaptation process and they propose a factor, "strength of belief in climate change," to explain the variance of adaptation at a local level. Based on a study conducted among individual private forest owners in Sweden, these researchers have inferred that the strength of belief in climate change and the adaptive capacity among individuals has a significant association with undertaking adaptation action.

2.4.2 Socio-Cognitive Model of Adaptation to Climate Change

To answer the question of what motivates some individuals to show adaptive behavior while others do not, Grothmann and Patt (2005) developed a socio-cognitive Model of Private Proactive Adaptation to Climate Change (MPPACC) based on Protection Motivation Theory (PMT). The model is set out in Figure 2.2.

The authors propose that risk perception and perceived adaptive capacity are two cognitive factors that determine a person's decision to drive the process of adaptation to climate change and its impacts. The process first starts with a risk appraisal, a person's perceived probability of being exposed to an adverse event or threat in the future, and the perceived consequences of that event to things that he or she values. This cognitive process of risk appraisal results in a particular risk perception that is influenced by the risk experienced in the past.

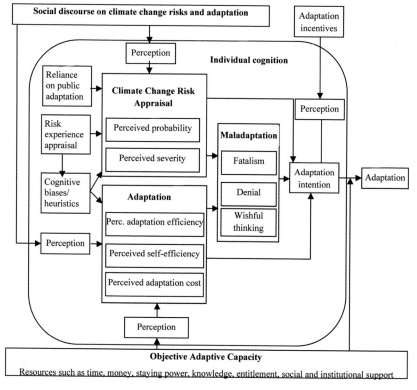

Figure 2.2 Process model of private proactive adaptation to climate change. *Source: Compiled from Grothmann and Patt (2005).*

This model, seen in Figure 2.2, suggests that a risk appraisal triggers on adaptation appraisal that includes a person's belief in the effectiveness of the adaptive response, the ability to perform adaptive actions and the assumed costs of implementing actions. People tend to assess their adaptive capacity when they believe that the risk has exceeded a specific threshold level. Burch and Robinson (2007) argue that the lay public's perception of risk is based on rational criteria, although these criteria are different from those of experts. These authors suggest that the perception of risk derives from a feeling of dread, i.e., the risk is assumed to be catastrophic and unknown (novel risks) and thus is strongly influenced by socio-economic factors. Thereby, individuals' constructs of reality can affect their perceived adaptive capacity and lead to an irrational judgment of risks. The MPPACC includes these as cognitive biases and heuristics.

According to the MPPACC, people's responses to climate risk, based on their risk and adaptive capacity appraisal processes, could be adaptation and/or maladaptation. A high perceived adaptive capacity and perceived climate risk result in adaptation action via the stage of adaptation intention with the positive influence of objective adaptive capacity and adaptation incentives (e.g., tax reduction).

The MPPACC model designed by Grothmann and Patt (2005) helps to describe and predict the adaptation process associated with climate change by analyzing the subjective dimension of adaptive capacity and climate risks in the context of objective adaptive capacity. This model demonstrates that socio-cognitive factors play a crucial role in linking adaptive capacity and adaptation action. Therefore, Kuruppu and Liverman (2011) suggest that cognitive barriers are to be addressed to facilitate adaptation in the communities by utilizing their adaptive capacity.

2.5 ADAPTATION TO CLIMATE CHANGE IN LIVELIHOOD FRAMEWORK

A livelihood "comprises the capabilities, assets and activities required for a means of living" (Scoones, 1998, p. 5). A livelihood is considered sustainable when it can cope with, and recover from, external shocks and stresses and can maintain the long-term productivity of the natural resource base while not undermining the livelihoods of others (DFID, 1999).

At the household and community level, the livelihood framework is considered to be helpful to explore how resource-dependent people in rural areas cope with the climate risk and uncertainty in their broader livelihood context (Badjeck et al., 2010). Basically, this approach focuses on assets that people use to build a satisfactory living in interaction with the contexts of vulnerability and transforming policy and institutions. Developed in the 1990s, by combining insights from farming system analysis, micro-economics, institutional analysis, and development economics, the livelihood approach has gained prominence in development work as a way of developing and applying unique frameworks by major development agencies such as UNDP, DFID, and CARE (AIACC, 2006). Among these frameworks, the Sustainable Livelihood Approach (SLA) has become the most prominent approach. It was developed by the UK Department for International Development (DFID) and aimed to alleviate poverty and promote livelihood oriented development (SDC, 2008).

2.5.1 Sustainable Livelihoods Approach (SLA)

The sustainable livelihoods approach (SLA) helps to organize various factors that constrain or improve how people make a living, and to understand how these factors correlate with each other. This approach recognizes that having access to certain assets, people put effort into making a living in the context of vulnerability and the prevailing social, institutional and organizational environment. The people apply livelihood strategies in these contexts to achieve beneficial livelihood outcomes, including higher material welfare, increased well-being, and reduced vulnerability. A schematic diagram of the sustainable livelihoods approach is shown in Figure 2.3.

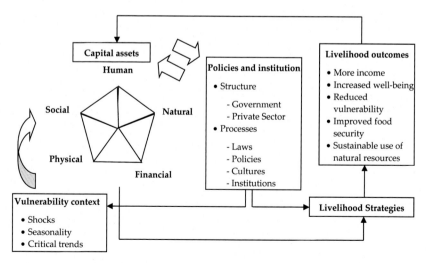

Figure 2.3 Sustainable livelihoods framework. *Source: Adapted from DFID (1999).*

In the livelihoods approach, three groups of components constitute the basic framework: (1) the asset portfolio forming the core elements of livelihoods; (2) the vulnerability context, and the structures and institutional processes; and (3) livelihood strategies and outcomes.

2.5.1.1 The Role of Assets

In the livelihoods approach, resources are regarded as assets or capital, and are often classified into the five types of assets drawn upon by household members for making their living: human capital (e.g., health, nutrition, education, knowledge, skills, capacity to work, capacity to adapt); social capital (e.g., networks, association, trust, shared values and behaviors, formal and informal groups, leaderships); natural capital (e.g., land, water, aquatic resources, forest, wildlife); physical capital (e.g., infrastructure, tools, and equipment); and financial capital (e.g., savings, credit and debt, remittances, pensions, wages) (Serrat, 2008). The livelihoods approach is fundamentally based on the understanding that people require a range of assets to achieve their beneficial livelihood outcomes because any single category of asset in its own right is not adequate to meet their varied livelihood objectives.

The access to these assets in the form of ownership, or the right to use in a given context, plays a key role in determining the vulnerability to the effects of external disturbances, including climate change (Ospina and Heeks, 2010). The increased access to a blend of assets results in a sustainable livelihood and stronger ability to respond to climate change. Thus, these livelihood assets act as a core element of both adaptive capacity and adaptation strategies.

2.5.1.2 Vulnerability Context

Vulnerability is characterized as an inability of individuals, households, and communities to avoid, cope with, or recover from the adverse impacts of factors that directly impact on their asset statuses and the strategies that are open to them to achieve beneficial livelihood outcomes. These factors are beyond their immediate control (DFID, 1999). The vulnerability context includes shocks (e.g., conflict, illness, floods, storms, drought), seasonalities (e.g., prices and employment opportunities), and critical trends (e.g., demographic, environmental, economic, governance, and technological trends) (Serrat, 2008; Malone, 2009).

2.5.1.3 The Role of Institution and Structures

The institution and structure constitute an important context of the livelihood framework, in the way that they effectively determine the access to various capital assets and livelihood strategies and influence government and private organizations; the terms of exchange between different types of capitals; and outcome of any livelihood strategies (DFID, 1999).

Institutions are structures of social order that shape political, economic, and social interactions and human agency (North, 1990). They can be formal or informal. Formal institutions are codified in legally building documents through governmental bureaucratic channels and are enforced by legal procedures such as laws, regulations, agreements, and operational arrangements (Pahl-Wostl, 2009). Informal institutions, on the other hand, are found in social or cultural norms and values, giving shape of the customary behavior of individuals and groups (Pelling and High, 2005). Structures, on the other hand, are public and private sector organizations that formulate and implement policies and legislations, deliver services, purchase, trade, and perform all manner of other functions that affect livelihoods (Serrat, 2008).

Serrat (2008) argues that organizations or agencies cannot be effective in the absence of appropriate institutions through which policies are formulated and implemented. Institutions thus play a key role in every aspect of livelihood, including granting or denying access to assets, affecting transformation of one type of asset into another, and providing incentives that stimulate better choices and dictate interpersonal relationships (DFID, 1999).

2.5.1.4 Livelihood Strategies and Outcomes

Livelihood strategies are a range of activities that people undertake to achieve their livelihood outcomes. Decisions on livelihood strategy may invoke multifarious repertoires of activities, mostly influenced by people's access to a level and combination of assets. Potential livelihood outcomes include more income, increased well-being, reduced vulnerability, improved food security, sustainable use of natural resource base, recovered human dignity, and so on (Serrat, 2008).

The livelihoods approach seeks to understand what factors motivate people's choice of livelihood strategies and where the major constraints lie to meeting their needs. An effectively performing institutional arrangement can expand the choices and flexibility of livelihood strategies by mitigating constraints and reinforcing the positive aspects as a way of facilitating the mobility in labor markets and reducing risks and transaction costs involved in embarking on new ventures (DFID, 1999).

2.5.2 Adapting Livelihood Approaches to Climate Change

Managing the risk associated with climate variability is integral to a comprehensive livelihood strategy of natural resource–dependent societies in developing countries, yet many are facing increasing pressure linked to global climate change and global economic change (Adger et al., 2007; Osbahr et al., 2008). The manifestations of climate change are slowly emerging in the increase in both the magnitude and frequency of extreme hazards, including flood and tropical cyclone. Long-term changes in climate trends, such as changes in rainfall and temperature, are thereby causing rapid and slow onset of disturbances in natural resources and ecosystem services on which poor communities depend for their subsistence and income. This has to aggravate the inherent vulnerability in low-income countries plagued by structural problems of poverty, underdevelopment, food and livelihood insecurity, socio-political inequalities, and power differentials (Tschakert and Dietrich, 2010).

Given this realization of the daunting challenges of climate change, adaptation to the impacts of climate change is now at the forefront of scientific research and policy negotiations (Tschakert and Dietrich. 2010). Although the global change community has experienced a recent surge in adaptation research, the underlying scientific understanding of the process of adaptation to climate change is still subject to theoretical and empirical challenges (Adger et al., 2007). In light of this challenge, this study investigates the components, processes, and characteristics that shape the adaptation of rural livelihoods facing climate variability, extreme events, and change.

Arnell (2010) proposes that the model of adaptation be constructed by taking account of local circumstances, encompassing the geophysical characteristics, governance and management practices, and institutional contexts that significantly affect the actual decision-making processes relating to adaptation. From this perspective, we use a sustainable livelihoods approach that provides a structure for organizing research on a practical adaptive response that is strongly grounded in local situations and involves participatory assessments. By extending the empirical coverage, this research adds critical mass to the evolving adaptation research focusing on specific contextual factors that actually drive or constrain adaptations on local or regional scales.

The study combines insights from literature pertaining to the sustainable livelihoods approach, together with the cognitive dimension of adaptation so as to develop a conceptual framework of the processes of livelihood adaptation to climate stressors. This proposed framework helps to explore how natural resource–dependent people in rural areas embed climate risk and uncertainty into their broader livelihood context to increase the resilience of rural households to climate-related shocks.

2.5.3 A Conceptual Framework for Livelihood Adaptation to Climate Variability and Change

This conceptual framework is an attempt to integrate climate change adaptation to DFID's Sustainable Livelihoods Approach (SLA), so as to understand and analyze the climate risk management processes at the household level in the vulnerable environment of developing economies. Figure 2.4 offers a schematic representation of the key features of the adaptation processes in the pursuit of resilient livelihoods under climatic uncertainty. The figure shows the relationship between the livelihood components, climate risk perceptions, adaptation actions, and livelihood outcomes.

The framework for the livelihood adaptation process illustrates four main process components, namely: (1) access to livelihood assets; (2) perception of climate risks; (3) embedding climate risk management in livelihood strategies; and (4) the institutional context that contributes to the adaptation process to achieve a climate-resilient livelihood. When used to holistically assess adaptation capacity in different cases, data would need to be collected for each component.

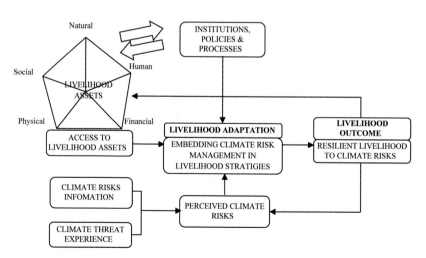

Figure 2.4 Conceptual framework for livelihood adaptation to climate variability and change. *Source: Adapted from DFID (1999).*

2.5.3.1 Access to Livelihood Assets

The access to the five core asset categories (human, social, natural, physical, and financial capitals) in the form of ownership or the right to use these assets is a primary concern of households in order to manage climate risk. Their capability to integrate climate risks is determined by a series of livelihood decisions that depend on a combination of household assets and the allocation of these assets in a given institutional and organizational environment to achieve meaningful livelihood outcomes (Tschakert and Dietrich, 2010). These capabilities are the household's adaptive capacity, which can be translated into their livelihood adaptations to reduce vulnerability to climate change and other stressors.

2.5.3.2 Climate Risk Perception

As has been discussed previously, adaptive capacity does not automatically translate into adaptive actions; the social and cognitive basis of the perception of climate risks triggers this process of adaptation. The socio-cognitive Model of Private Proactive Adaptation to Climate Change (MPACC) demonstrates that people's previous experience with threats (hazards) and the social discourses on climate change construct the perception of future climate risks, which initiates a cascading process of adaptation mediated by the wider livelihood context. The adaptations can be both reactive, that is, undertaken after experiencing the impacts of climatic events, and proactive, whereby initiatives are taken to avoid future damages.

2.5.3.3 Embedding Climate Risk Management into Livelihood Strategies

The SLA suggests that the adaptation process is part of the livelihood strategies that people undertake to respond to climatic risk within the wider livelihood context (Ospina and Heeks, 2010). Triggered by climate risk perceptions, adaptation processes can lead to an incremental adjustment of a livelihood system, manifest by embedding climate-related risk management in the routine livelihood activities and outcomes to withstand and recover from short-term, weather-related shocks and also changes in the long-term climate-related trends. The adaptation processes also involve system transformations when new livelihood strategies are adopted in response to climate threat, as well as taking opportunities that stem from climate change.

2.5.3.4 Institutional Context

Social, political, institutional, and organizational arrangement provide contexts in which people construct and adapt their livelihood strategies (DFID, 1999). The adaptations are seldom isolated activities; rather, they are an ongoing process in which people draw on their assets in the institutional framework and pursue different livelihood adaptations to climate variability and changes at different time scales that ultimately contribute to achieve their beneficial livelihood outcomes. In this respect, government policies and institutional frameworks regarding both climate and nonclimate issues play a crucial role in facilitating or hindering people's response strategies to manage climate risks.

2.5.3.5 Climate-Resilient Livelihood Outcomes

Livelihood outcomes that are resilient to climate shocks are the goals of households that these households pursue through the investment of different combinations of livelihood assets in livelihood strategies embedded with climate risk management strategies. Livelihood adaptation analysis, therefore, should be focused on exploring whether the adaptation actions contribute to the achievement of livelihood outcomes that are priorities for local people (e.g., reduced vulnerability, increased well-being, and increased income). The proposed adaptation framework is intended to gain an understanding of the causality in rural livelihood systems.

2.5.3.6 Feedback Mechanisms

There are important feedback relationships in the livelihood systems that affect adaptation processes. Adaptation strategies that generate beneficial livelihood outcomes have implications for reducing vulnerability and increasing people's access to a blend of capital assets, and gradually build adaptive capacity. Reduced vulnerability, in turn, alters the people's risk perceptions of climate variability and change.

The conceptual framework (Figure 2.4) has been developed to guide the study. This analytical framework allows an identification of the main components, their relationships, and feedback mechanisms that affect the livelihood adaptation process at a household level. However, this study has a limited scope, as it is restricted to the research questions in Section 1.2, and therefore focuses primarily on the relevant elements of the framework.

REFERENCES

Adger, W.N., Agrawala, S., Mirza, M.M.Q., Conde, C., O'Brien, K., Pulhin, J., Pulwarty, R., Smit, B., Takahashi, K., 2007. Assessment of adaptation practices, options, constraints and capacity. In: Parry, M.L., Canziani, O.F., Palutikof, J.P., van der Linden, P.J., Hanson, C.E. (Eds.), IPCC Climate Change 2007: Impacts, Adaptation and Vulnerability. Contribution of Working Group II to the Fourth Assessment Report of the Intergovernmental Panel on Climate Change. Cambridge University Press, Cambridge, pp. 717–743.
Adger, W.N., 2006. Vulnerability. Global Environ. Change 16, 268–281.

Adger, W.N., Brown, K., Nelson, D.R., Berkes, F., Eakin, H., Folke, C., Galvin, K., Gunderson, L., Goulden, M., O'Brien, K., Ruitenbeek, J., Tompkins, E.L., 2011. Resilience implication of policy response to climate change. WIREs Clim. Change 2, 757–766.

AIACC, 2006. Vulnerability and Adaptation to Climate Variability and Change: The Case of Farmers in Mexico and Argentina. Assessment of Impacts and Adaptations to Climate Change (AIACC). Project No. LA-29. The International START Secretariat, USA.

Arnell, N.W., 2010. Adapting to climate change: an evolving research programme. Clim. Change 100, 107–111.

Berrang-Ford, L., Ford, D.J., Paterson, J., 2011. Are we adapting to climate change? Global Environ. Change 21, 25–33.

Berkhout, F., Hertin, J., Gann, D., 2004. Learning to Adapt: Organizational Adaptation to Climate Change Impacts. Tyndall Centre Technical Report No.11. Tyndall Centre for Climate Change Research, UK.

Berkes, F., Colding, J., Folke, C., 2003. Introduction. In: Berkes, F., Colding, J., Folke, C. (Eds.), Navigating Social-Ecological Systems: Building Resilience for Complexity and Change. Cambridge University Press, Cambridge, UK, pp. 1–29.

Berkes, F., 2007. Understanding uncertainty and reducing vulnerability: lessons from resilience thinking. Natural Hazards 41, 283–295.

Barnett, J., O'Neill, S., 2010. Maladaptation. Global Environ. Change 20, 211–213.

Blennow, K., Persson, J., 2009. Climate change: motivation for taking measures to adapt. Global Environ. Change 19, 100–104.

Burch, S., Robinson, 2007. A framework for explaining the links between capacity and action in response to global climate change. Clim. Policy 7, 304–316.

Badjeck, M.C., Allison, E.H., Halls, A.H., Dulvy, N.K., 2010. Impacts of climate variability and change on fishery-based livelihoods. Mar. Policy 34, 375–383.

Cubasch, U., Wuebbles, D., Chen, D., Facchini, M.C., Frame, D., Mahowald, N., Winther, J.-G., 2013. Climate change 2013: the physical science basis (Introduction). In: Stocker, T.F., Qin, D., Plattner, G.-K., Tignor, M., Allen, S.K., Boschung, J., Nauels, A., Xia, Y., Bex, V., Midgley, P.M. (Eds.), Contribution of Working Group I to the Fifth Assessment Report of the Intergovernmental Panel on Climate Change. Cambridge University Press, Cambridge, United Kingdom and New York, NY, USA.

Cash, D.W., Adger, W.N., Berkes, F., Garden, P., Lebel, L., Olsson, P., Pritchard, L., Yound, O., 2006. Scale and cross-scale dynamics: governance and information in a multilevel world. Ecol. Soc. 11 (2), 8.

DFID, 1999. Sustainable Livelihoods Guidance Sheets. Department for International Development (DFID), Norwich, UK.

Encyclopedia Britannica, 2011. http://www.britannica.com/Ebchecked/topic/5263/adaptation (accessed 18.10.11.).

Grothmann, T., Patt, A., 2005. Adaptive capacity and human cognition: the process of individual adaptation to climate change. Global Environ. Change 15, 199–213.

Hinkel, J., 2011. "Indicators of vulnerability and adaptive capacity": towards a clarification of the science-policy interface. Global Environ. Change 21, 198–208.

Ionescu, C., Klein, R., Hinkel, J., Kumar, K., Klein, R., 2009. Towards a Formal Framework of Vulnerability to Climate Change. Environ. Model Assess 14, 1–16.

IPCC, 2012. Summary for policymakers: managing the risks of extreme events and disasters to Advance climate change adaptation. In: Field, C.B., Barros, V., Stocker, T.F., Qin, D., Dokken, D.J., Ebi, K.L., Mastrandrea, M.D., Mach, K.J., Plattner, G.K., Allen, S.K., Tignor, M., Midgley, P.M. (Eds.). A Special Report of Working Groups I and II of the Intergovernmental Panel on Climate Change. Cambridge University Press, Cambridge, UK, and New York, NY, USA, pp. 3–21.

IPCC, 2001. Climate Change 2001: Impacts, Adaptation, and Vulnerability. Contribution of Working Group II to the Third Assessment Report of the Intergovernmental Panel on Climate Change. Cambridge University Press, Cambridge, UK.

Kuruppu, N., Liverman, D., 2011. Mental preparation for climate adaptation: the role of cognition and culture in enhancing adaptive capacity of water management in Kiribati. Global Environ. Change 21, 657–669.

Malone, E.L., 2009. Vulnerability and Resilience in the Face of Climate Change: Current Research and Needs for Population Information. Battelle Memorial Institute, Washington.

Nelson, D.R., 2011. Adaptation and resilience: responding to changing climate. WIREs Clim. Change 2, 113–120.

Naess, L.A., Bang, G., Eriksen, S., Vevatne, J., 2005. Institutional adaptation to environmental change: flood responses at the municipal level in Norway. Global Environ. Change 15, 125–138.

North, D., 1990. Institutions, Institutional Change and Economic Performance. Cambridge University Press, UK.

Osbahr, H., Twyman, C., Adger, W.N., Thomas, D.S.G., 2008. Effective livelihood adaptation to climate change disturbance: scale dimensions of practice in Mozambique. Geoforum 39, 1951–1964.

Orlove, B., 2009. The past, the present and some possible futures of adaptation. In: Adger, N.W., Lorenzoni, I., O'Brien, K.L. (Eds.), Adapting to Climate Change: Thresholds, Values, Governance. Cambridge University Press, New York, p. 132.

Ospina, A.V., Heeks, R., 2010. Linking ICTs and Climate Change Adaptation: A Conceptual Framework for e-Resilience and e-Adaptation. Centre for Development Informatics, Institute for Development Policy and Management, University of Manchester, UK.

Parry, M., Lowe, J., Hanson, C., 2009. Overshoot, adapt and recover. Nature 458, 1102–1103.

POST, 2006. Food Securities in Developing Countries. Parliamentary Office of Science and Technology, UK.

Pahl-Wostl, C., 2009. A conceptual framework for analysing capacity and multi-level learning process in resource governance regimes. Global Environ. Change 19, 354–365.

Pelling, M., High, C., 2005. Social Learning and Adaptation to Climate Change. Disaster Studies Working Paper 11. Benfield Hazard Research Centre, University College London, UK.

Smith, J.B., Schneider, S.H., Oppenheimer, M., Yohee, G.W., Haref, W., Mastrandrea, M.D., Patwardhan, A., Burton, I., Corfee-Morlot, J., Magadza, C.H.D., Fussel, H.M., Pittock, A.B., Rahman, A., Suarez, A., Ypersele, J.P., 2009. Assessing dangerous climate change through an update of the Intergovernmental Panel on Climate Change (IPCC) "reason for concern". Proc. Natl. Acad. Sci. U.S.A. 106, 4133–4137.

Smit, B., Pilifosova, O., Burton, I., Challenger, B., Huq, S., Klein, R.J.T., Wandel, J., 2000. An anatomy of adaptation to climate change and variability. Clim. Change 45, 223–251.

Smit, B., Pilifosova, O., Burton, I., Challenger, B., Huq, S., Klein, R.J.T., Yohe, G., 2001. Adaptation to climate change in the context of sustainable development and equity. In: McCarthy, J.J., Canziani, O., Leary, N.A., Dokken, D.J., White, K.S. (Eds.), Climate Change 2001: Impacts, Adaptation and Vulnerability. Contribution of the Working Group II to the Third Assessment Report of the Intergovernmental Panel on Climate Change. Cambridge University Press, Cambridge, pp. 877–912.

Smithers, J., Smit, B., 2009. Human adaptation to climatic variability and change. In: Schipper, E.L.F., Burton, I. (Eds.), The Earthscan Reader on Adaptation to Climate Change. Earthscan, UK.

Smit, B., Skinner, M.W., 2002. Adaptation options in agriculture to climate change: a typology. Mitigation Adapt. Strategies Global Change 7, 85–114.

Scoones, I., 1998. Sustainable Livelihoods: A Framework for Analysis. Working Paper 72. Institute for Development Studies, Brighton, UK.

SDC, 2008. Putting a Livelihood Perspective into Practice. Swiss Agency for Development and Cooperation, Switzerland.

Serrat, O., 2008. The Sustainable Livelihoods Approach. Knowledge Solutions, Asian Development Bank.

Thomas, R.J., 2008. Opportunity to reduce vulnerability of dryland farmers in Central and West Asia and North Africa to climate change. Agric. Ecosyst. Environ. 1, 71–90.

Tschakert, P., Dietrich, K.A., 2010. Anticipatory learning for climate change adaptation and resilience. Ecol. Soc. 15 (2), 11.

United Nations, 1992. The United Nations Framework Convention on Climate Change.

World Bank, 2010. World Development Report 2010: Development and Climate Change. World Bank, Washington, DC.

Walker, B., Holling, C.S., Carpenter, S.R., Kinzig, A., 2004. Resilience, adaptability and transformability in social-ecological systems. Ecol. Soc. 9 (2), 5.

Wesche, S., Armitage, D.R., 2010. From the inside out: a multi-scale analysis of adaptive capacity in a Northern community and the governance implications. In: Armitage, D.R., Plummer, R. (Eds.), Adaptive Capacity and Environmental Governance. Spinger-Verlag, Dordrecht, Germany.

CHAPTER 3

Study Design and Data Sources

Contents

3.1 Introduction	29
3.2 Selection of Field Research Methods	30
3.3 Study Design	31
3.3.1 Climate Variability and Extremes as a Proxy for Climate Change	31
3.3.2 Sustainable Livelihoods Approach: Evaluation Framework for Livelihood Adaptation to Climate Change	32
3.3.3 Institutional Analysis Framework	33
3.4 Case Study	34
3.4.1 Scoping the Case Study Area	34
3.4.2 Selection of Case Study	35
3.4.3 Selection of Communities	35
3.4.4 Selection of Sample Households	35
3.5 Field Data Collection	36
3.5.1 Secondary Information Review	36
3.5.2 Introductory Community Visit	36
3.5.3 Household Survey	37
3.5.4 Focus Group	37
3.5.4.1 Hazard Mapping	*37*
3.5.4.2 Seasonal Calendar	*38*
3.5.5 Key Informant Interview	38
3.5.6 Climate Data Collection	38
References	39

3.1 INTRODUCTION

This chapter discusses the research approaches used to collect and analyze data to assess the processes of livelihood adaptation to climate variability and change as social resilience in the coastal region of Bangladesh. The first section of this chapter starts with an explanation of the rationale behind using a mixed study design for the research plan, which integrates qualitative and quantitative methods for data collection and analysis in the field of climate change adaptation. Drawing on the relevant literature, this section presents the strengths of a mixed-method approach as a strategic research methodology over qualitative or quantitative research methods. This is followed by a discussion on the study design and data collection methods to be used in

Experiencing Climate Change in Bangladesh
http://dx.doi.org/10.1016/B978-0-12-803404-0.00003-X

this field research to evaluate the adaptation situation in its natural settings. As a logical sequence, in the next section we explain the techniques used for analyzing both the qualitative and quantitative data and link these two types of data. The final section describes the strategies used for validating and verifying the data and the results of analysis.

3.2 SELECTION OF FIELD RESEARCH METHODS

Within the social and behavioral sciences, there are basically two kinds of research traditions—namely, qualitative and quantitative research methods. These two traditions constitute the core of the general methodological scheme of research (Valsiner, 2000). The quantitative method is guided by positivist paradigmatic approaches to research. This paradigmatic view of the wold assumes that social reality has an objective ontological structure that can be measured and explained scientifically, in much the same way as in the natural sciences. This approach centers on formulating hypotheses and verifying them deductively. The quantitative approach involves a measuring of the events succinctly; performing statistical analysis of numerical data; and making summaries, comparisons and generalizations from statistics with a certain precision. One of the major limitations of the quantitative approach is that it fails to provide information about the "real-world" context in which the studied phenomenon occurs (Moghaddam et al., 2003).

In contrast, the ontological assumption of qualitative research is that social reality is constructed and sustained through the subjective experience of people involved in communication, such as in language, consciousness, shared meanings, documents, tools, and other artifacts (Morgan and Smircich, 1980; Klein and Myers, 1999). Qualitative methods are particularly oriented toward accurately describing, decoding, and interpreting the meaning of phenomena happening in the normal social context (Fryer, 1991). This approach allows the researcher to study the selected issue, event, or case in depth and detail within its original context (Patton, 1987). The weaknesses of the qualitative method include the following: (1) reliability in assessing the links and association among observations, cases, or constructs may be lacking (Castro et al., 2010); (2) it is difficult to arrive at definitive conclusions because this approach may bias toward the personal characteristics of the researchers; and (3) focusing on in-depth analysis of a small sample sizes limits the method to producing generalized findings (Giddens, 1984).

Mixed-method research is a pragmatic research approach that has emerged to draw on the strengths and minimize the weaknesses of the qualitative and quantitative research traditions. Johnson and Onwuegbuzie (2004, p. 17) define mixed methods research as "the class of research where the researcher mixes or combines quantitative and qualitative research techniques, methods, approaches, concepts or language into a single study." They argue that researchers in today's interdisciplinary, complex, and dynamic research environment need to complement qualitative methods with quantitative ones, and, in effect, the mixed method offers the best opportunities for answering important research questions.

Greene et al. (1989, cited in Johnson and Onwuegbuzie, 2004) noted five major rationales for conducting mixed methods research: (1) triangulation, which seeks a convergence and corroboration of results from different methods studying the same phenomenon; (2) complementarity, in which results from one method are used for clarification with results from the other method; (3) elaboration, as in focusing on the depth and detail of the research by using different methods for different inquiry components; (4) initiation, which stimulates recasting the research question as a result of discovering paradoxes and contradictions; and (5) development (i.e., using the findings from one method to help inform the other method).

Miller et al. (2010) note that researchers working at the society–nature interface, such as in resilience and vulnerability research, are increasingly using mixed methodologies because of the growing awareness of the integrated nature of the problems under study. They argue that a combination of qualitative and quantitative research methods produces more success than any of the single methods in covering the context of the most vulnerable groups in society. This has been supported by Ziervogel et al. (2006), who note that a mixed-method approach provided a more holistic depiction of the environmental and socio-economic stresses of individual actions and decision-making processes when they empirically studied the local adaptation strategies to changes in climate variability. With these in mind, this study aligns itself with a mixed-method research approach.

3.3 STUDY DESIGN

The following elements comprise the conceptual and technical design of this research project:

3.3.1 Climate Variability and Extremes as a Proxy for Climate Change

Despite the considerable progress in understanding how the climate system works, there is still scientific uncertainty about the future climate as well as uncertainty about evolving socio-political–economic system to confidently assess the impacts of climate change (Yohe and Oppenheimer, 2011). Considering the current limitations of knowledge, this research project will be based on the understanding that excessive vulnerability to current climate variability and extremes is a good proxy for future vulnerability under climate change (Heltberg and Bonch-Osmolovskiy, 2011), which is very much relevant to the case of Bangladesh, which has been identified as a country extremely vulnerable to climate change. Siegfried (2005 cited in AIACC, 2006) notes that the conditions and processes that foster a coping and adaptive capacity to today's climate variability and extremes provides a useful indication as to the likely resiliency of the system to future climate change. With this assumption, we investigate the households' adaptation ability, in relation to current climatic disturbances that have a great influence on their ability to adapt to anticipated climate change.

3.3.2 Sustainable Livelihoods Approach: Evaluation Framework for Livelihood Adaptation to Climate Change

As discussed in Section 2.5 the livelihoods framework provides an analytical structure that is useful to understand the nuances of coping and adaptive strategies that individuals undertake to respond to climate risk (AIACC, 2006; Thomas et al., 2007). The notion of the five capitals in livelihood models captures the concept of adaptive capacity of households and individuals that shapes the adaptation processes and options. The livelihood framework also provides broader picture that helps to understand the people's perceptions about climate variability and change, which is considered to be a trigger to behavioral responses. Taking into account the above-mentioned benefits of using the livelihood approach, we used our own variant of a livelihood framework, mostly built on the DFID model of sustainable livelihood framework (a framework of livelihood adaptation process to climate change, outlined in the Section 2.5.2). This was to structure our investigation by sequencing it, first, to investigate the livelihood features encompassing the vulnerability context, households' assets, and strategies; second, people's perceptions of, and responses to, changes in climate variability and extremes; and, finally, the role of policies and institutions in local adaptations to climate change, focusing on the strategies of embedding climate risk with livelihood activities in pursuit of securing livelihood outcomes that are resilient to climate disturbances.

Table 3.1 provides variables selected to represent each of the five livelihood capitals that feature the main resources available to households for adaptation practices to reduce vulnerability and enhance resilience to climatic

Table 3.1 A list of variables representing the five livelihood capitals influencing the adaptive practices of rural households in the coastal area

Capital	Variable	Description
Human	Household size	Number of household member
	Dependency ratio	Ratio of dependents to working household members
	Education	Number of years in school completed
	Health	Self-assessed health status
	Technical training	Training of household members to manage livelihood practices
Physical	Housing	Condition of dwelling structure
	Equipment	Possession of equipment for economic production
	Access to safe water	Sources of water used by household (access to safe water)
	Access to cyclone shelter	Number of cyclone shelters in the locality for the community
Natural	Land	Total amount of land operated by the household
	Livestock	Total number of ruminants and poultry
Financial	Income	Total cash income of the household from the most important livelihood activities in 1 year (year preceding the survey)
	Access to finance	Borrowing capacity of the household
Social	Community participation	Membership of family members in community organizations
	Contact with NGO	Getting support from contact with NGOs

NGO, nongovernmental organization.

variability and change. These variables are often used to evaluate the adaptive capacity to climate change and livelihood assessment of the rural communities (DFID, 1999; AIACC, 2006; Nelson et al., 2007). To be relevant to a local setting, a provisional list of variables was prepared, based on the literature review, and finalized by consultation with local communities.

3.3.3 Institutional Analysis Framework

This investigation into the role of institutions in livelihood adaptation to climate change is framed to enhance an understanding of the functions of the formal institutions and their significance in supporting livelihood adaptive actions to manage climate variability and change. In examining the role of the institutions in shaping livelihood responses to climate disturbances, the study focuses on three approaches: social protection (SP), disaster risk reduction (DRR), and climate change adaptation (CCA). These dominant approaches overlap considerably with the broader objectives, policies, and practices contributing to reducing vulnerability to climate variability and extreme events and their impacts on livelihoods (Davies et al., 2009).

By focusing on these approaches, we systematically analyze the elements of the SP, DRR, and CCA policies and practices to understand how they mediate the livelihood adaptation processes at the local level. Within these three fields, there are numerous public organizations and nongovernmental organizations (NGOs) functioning at different administrative levels, ranging from the micro to the national, which are associated with various interventions. Of these various institutions, the study focuses on the policies, practices, and stakeholders currently involved in facilitating climate-resilient livelihoods at the local level, i.e., the coastal area of Bangladesh.

To investigate the extent to which policy affects people's ability to deal with climate change, we begin by noting the livelihood adaptation issues that were considered crucial at a household and community level. From the specific issues raised by local communities, only the larger issues that are sufficiently representative of the wider problems in supporting households' response to climate change in the study region were addressed. With this local point of view, the policies and practices of the governmental and nongovernmental agencies involved in addressing key issues and problems related to climatic disturbances in the coastal area were reviewed. In view of the local definitions of the adaptation issues, the study relates the issues from the field to the national institutional context, where policy is formulated and enshrined in legislation.

3.4 CASE STUDY

A case study in Bangladesh was selected to apply the research methodology so as to explore the process by which rural households in the coastal area adapt to climate variability and change. The case study also explored the roles of the agencies and formal institutions in supporting livelihood responses to climate change at a local level.

3.4.1 Scoping the Case Study Area

The case study area was selected to ensure that it covers the distinctive biophysical environment of the coastal area of Bangladesh. It also needed to adequately represent the hydro–climatic events of a coastal area manifested by regular tropical cyclones, flooding, and salt water intrusion; as well as an increased frequency/intensity of these climate extremes in the future under a climate change scenario; and, finally, the livelihood systems of the coastal region of Bangladesh.

3.4.2 Selection of Case Study

Having met these basic criteria, the Mongla Sub-district in the Bagerhat District that lies in the southwest coastal region of Bangladesh was chosen to conduct field research. Chapter 4 gives an overview of the study region.

3.4.3 Selection of Communities

A union of Mongla named Chila was selected, with the assistance of the local NGOs. A union is the lowest administrative unit of government, consisting of several villages. The villages in Chila Union are geographically linked to each other through a common border and are socially linked through relationship. Moreover, they share the same livelihood systems and vulnerability context. Therefore, Chila was considered as a single case.

3.4.4 Selection of Sample Households

The next stage involved the selection of sample households from the communities. The population universe for this study was based on the list of households available from the Chila Union Council in Mongla Sub-district. After having the final list of the target population, a probability sampling design was chosen because it is more objective and is closely associated with scientific research (McGrew and Monroe, 2000). The sample size was determined by using the following formula (Sampling and Surveying Handbook, 2002):

$$n = \frac{NZ^2 * 0.25}{[d^2 * (N-1)] + (Z^2 * 0.25)}$$

where

n = sample size required

N = total population size

d = precision level

Z = number of standard deviation units of the sampling distribution corresponding to the desired confidence level

To minimize the risk relating to sample size determination, it is considered appropriate to choose a 95% confidence level and ±5% precision

Table 3.2 Categorization of household based on land holding size

Type of household	Size of farm holding (acres)
Nonfarm	<0.05
Marginal farm	0.05–0.49
Small farm	0.50–2.49
Medium farm	2.5–7.49
Large farm	≥7.50

1 acre = 0.40 ha.
Source: Report on Metadata for National Agricultural Statistics in Bangladesh 2007.

level ($d = 0.05$, $Z = 1.95$). In 2012, the total households in Chila were 4794; then:

$$n = \frac{4794 * 1.96^2 * 0.25}{[0.05^2 * (4794 - 1)] + (1.96^2 * 0.25)} = 355.73$$

So, the required sample size was 356, as computed above. However, to further minimize the risk of drawing inference about the population on the basis of sample information, the total sample size was increased to 372. The sample was then stratified by land holding size, where the size of each group was approximately proportional to the relative size of that group in the total population in the survey villages. Based on the size of a farm holding, households are classified in Bangladesh as nonfarm, marginal, small, medium, and large. The classification is shown in Table 3.2:

This stratified sampling design is considered to yield the most precise results because of the fact that people who belong to different livelihood groups have variable adaptation capacities, and in the rural setting these generally correspond to the farm size of the household.

3.5 FIELD DATA COLLECTION

3.5.1 Secondary Information Review

For climate change adaptation policy analysis, extensive documentary records of the policies, legislation, strategies, plans, project documents, assessment reports, and so forth were collected, examined, and summarized.

3.5.2 Introductory Community Visit

The initial site visit was intended as my introduction to the local community, to identify and contact local informants and to schedule subsequent fieldwork. At this stage, we synthesized information obtained from secondary sources and worked with key informants to cross-check that information.

3.5.3 Household Survey

The study used an explanatory sequential design (Creswell, 2009), which involved a first phase of quantitative data collection and analysis. The data gathering at this stage was carried out by using a semistructured questionnaire. The questionnaire was designed in the form of an interview schedule to generate information covering the demographic patterns, source of income, land use, livelihood assets, farming practices, social network, livelihoods disturbances, and responses of the households. In this research, the household is the basic unit of analysis, which is defined as a group of people living in the same residence, and who contribute food or income to the unit (Osbahr et al., 2010).

The initial findings of the household survey, and the group discussions with a cross-section of the livelihood groups, made it clear that shrimp farming was causing rapid land-use changes in the study area, and thus illustrated that the livelihoods of most villagers were intricately involved in this sector. This led us to seek an extended and deeper understanding of the important socio-economic and environmental aspects relating to shrimp farming as well as the shrimp farmers' ($n = 30$) accounts of the climate risks and other stressors affecting shrimp farming, along with the specific actions that shrimp farmers take as a response.

Following the data gathering and the preliminary analysis in the first phase, participatory techniques such as focus group discussions and key informant interviews were used to gain an insight into the livelihood dynamics, perception of risks and the impacts on the livelihood systems, current copping strategies, and how decisions are taken in their livelihood strategies in response to the changing environment, including the climate.

3.5.4 Focus Group

Focus groups were used for eliciting detailed information on specific issues by employing a wide range of techniques, including participatory approaches. Having acquired background information on the community, we selected a total of 20 focus groups, each with five to eight members, with the assistance of local facilitators.

3.5.4.1 Hazard Mapping

This was a mapping exercise in which the community maps the location of important livelihood resources and resources at risk from climate hazard in different locations. Hazard mapping formed the foundation for a proper understanding of the people's perceptions of changes in the climate in the recent past, in terms of the variability and predictability in different climate characteristics, such as rainfall, dry conditions, floods, and cyclone. This mapping exercise also helped to understand the local perceptions of the risks associated with climate and of their impacts on the livelihoods pursued by different groups within the community.

3.5.4.2 Seasonal Calendar

It is a useful tool to understand regular events that occur cyclically in a community every year. A seasonal calendar was designed by participants to list important events, such as food scarcity, migration, extreme events/disasters, and seasonal illness, and then the corresponding timing of the events were plotted. This exercise also included the development of an agricultural calendar in which the timing of agricultural activities were plotted along with the likely climatic conditions.

3.5.5 Key Informant Interview

Interviews with key informants were conducted at a village level in an unstructured format, as these meetings were intended for triangulation and cross-checking of the findings.

3.5.6 Climate Data Collection

We accessed the monthly precipitation and the maximum and minimum temperature data from Mongla meteorological station. The local temperature and rainfall data, spanning the period of about two decades were graphed (Figures 6.1 and 6.2) to visualize the climatic conditions to which the people in Mongla have recently been exposed. The mean of the total length of observation and the 5-year running mean were used to evaluate the extent to which local perceptions of climate change correlate with recent climate variability. Given the shorter period of records obtained from Mongla station, the observed trends are more representative of short-term variability rather than long-term climate change. To obtain scientifically tested evidence of changes in climate in the study area, we therefore relied on the findings of published literature in which researchers have analyzed the long-term (>50 years) temperature and rainfall data from the adjacent meteorological stations.

The time series on cyclone data used here came from the Water Resources Planning Organization of Bangladesh (WARPO) Web site (http://www.warpo.gov.bd/rep/knowledge_port/KPED/Process/Wind_StormWaves/Table Cyclone_History.htm) for the period 1960–1991. As the WARPO data were not up-to-date, we relied on the Emergency Events Database (EM-DAT) of the Centre for Research on the Epidemiology of Disasters (CRED), Belgium (EMDAT) (Web site: http://www.emdat.be/disaster-list) for the period between 1992 and 2013.

REFERENCES

AIACC, 2006.Vulnerability and Adaptation to Climate Variability and Change:The Case of Farmers in Mexico and Argentina. Assessment of Impacts and Adaptations to Climate Change (AIACC). Project No. LA-29.The International START Secretariat, USA.

Castro, F.G., Kellison, J.G., Boyd, S.J., Kopak, A., 2010. A methodology for conducting integrative mixed methods research and data analyses. J. Mixed Methods Res. 4 (4), 342–360.

Creswell, J.W., 2009. Research Design: Qualitative, Quantitative, and Mixed Methods Approaches. Sage, California.

DFID, 1999. Sustainable Livelihoods Guidance Sheets. Department for International Development (DFID), Norwich, UK.

Davies, M., Oswald, K., Mitchell, T., 2009. Climate change adaptation, disaster risk reduction and social Protection. In: Promoting Pro-Poor Growth: Social Protection, Organization for Economic Cooperation and Development, pp. 201–217.

Fryer, D., 1991. Qualitative methods in occupational psychology: reflections upon why they are so useful but so little used. Occup. Psychol. 14, 3–6. Special issue on qualitative methods.

Giddens, A., 1984.The Construction of the Society: Outline of the Theory of Structuration. Polity Press, Cambridge, UK.

Heltberg, R., Bonch-Osmolovskiy, M., 2011. Mapping Vulnerability to Climate Change. Policy Research Working Paper 5554.The World Bank.

Johnson, R.B., Onwuegbuzie, A.J., 2004. Mixed methods research: a research paradigm whose time has come. Educ. Res. 33 (7), 14–26.

Klein, H.K., Myers, M.D., 1999. A set of principles for conducting and evaluating interpretive field studies in information systems. Manage. Inf. Syst. Q. 12, 67–88.

Moghaddam, F.M., Walker, B.R., Harre, R., 2003. Cultural distance, levels of abstraction, and the advantages of mixed methods. In: Tashakkori, A., Teddlie, C. (Eds.), Handbook of Mixed Methods in Social & Behavioral Research. SAGE Publications, Thousand Oaks, CA, pp. 111–134.

Morgan, G., Smircich, L., 1980. The case for qualitative research. Acad. Manage. Rev. 5 (4), 491–500.

Miller, F., Osbahr, H., Boyd, E., Thomalla, F., Bharwani, S., Ziervogel, G., Walker, B., Birkmann, J., Vander Leeuw, S., Rockström, J., Hinkel, J., Downing, T., Folke, C., Nelson, D., 2010. Resilience and vulnerability: complementary or conflicting concepts? Ecol. Soc. 15 (3), 11.

McGrew, J.C., Monroe, B.C., 2000. An Introduction to Statistical Problem Solving in Geography. McGraw-Hill Higher Education.

Nelson, D.R., Adger, N.W., Brown, K., 2007. Adaptation to environmental change: contributions of a resilience framework. Annu. Rev. Environ. Resour. 32, 395–419.

Osbahr, H., Twyman, C., Adger, W.N., Thomas, D.S.G., 2010. Evaluating Successful Livelihood Adaptation to Climate Variability and Change in Sothern Africa. Ecol. Soc. 15 (2), 27.

Patton, M.Q., 1987. How to Use Qualitative Methods. SAGE Publications, Newbury Park, California. pp. 7–22.

Sampling and Surveying Handbook, 2002. Air University Sampling and Surveying Handbook: Guidelines for planning, organizing and conducting surveys. Air University, USA.

Thomas, D.S.G., Twyman, C., Osbahr, H., Hewitson, B., 2007. Adaptation to climate change and variability: farmer responses to intra-seasonal precipitation trends in South Africa. Clim. Change 83, 301–322.

Valsiner, J., 2000. Data as Representations: Contextualizing Qualitative and Quantitative Research Strategies. Social Science Information. SAGE Publication. pp. 99–113.

Yohe, G., Oppenheimer, M., 2011. Evaluation, characterization, and communication of uncertainty by the intergovernmental panel on climate change—an introductory essay. Clim. Change 108, 629–639.

Ziervogel, G., Bharwani, S., Downing, T.E., 2006. Adapting to climate variability: pumpkins, people and pumps. Nat. Resour. Forum 30, 294–305.

CHAPTER 4

The Research Setting

Contents

4.1	Introduction	42
4.2	Coastal Bangladesh	43
	4.2.1 Physical Setting	43
	4.2.2 Socio-Economic Context	44
	4.2.2.1 *Population*	*44*
	4.2.2.2 *Poverty and Unemployment*	*45*
	4.2.2.3 *Livelihood Characteristics*	*45*
	4.2.2.4 *Agriculture and Structure of Land Distribution*	*45*
	4.2.2.5 *Shrimp Farming*	*46*
	4.2.2.6 *Fishing*	*47*
	4.2.2.7 *Institutional Environment*	*48*
	4.2.2.8 *Role of NGOs*	*48*
	4.2.3 Characteristics of the Climate	48
	4.2.4 Biophysical Hazards and Key Impacts	50
	4.2.4.1 *Waterlogging and Drainage Congestions*	*50*
	4.2.4.2 *Salinity Intrusion*	*51*
	4.2.4.3 *Tropical Cyclone and Storm Surge*	*51*
	4.2.5 Potential Impacts of Climate Change	53
4.3	Description of the Study Area	53
	4.3.1 Geography of the Study Area	55
	4.3.2 Socio-Economic Features	55
	4.3.2.1 *Population Dynamics*	*55*
	4.3.2.2 *Health and Education Facilitates*	*56*
	4.3.2.3 *Other Public Services*	*57*
	4.3.2.4 *Road Infrastructure*	*57*
	4.3.2.5 *Livelihood Activities*	*57*
	4.3.2.6 *Farming*	*57*
	4.3.2.7 *Fisheries*	*60*
	4.3.2.8 *Agriculture*	*61*
	4.3.2.9 *Extraction of Sundarbans Forest Resources*	*61*
	4.3.2.10 *Seasonal Migration*	*62*
	4.3.3 Current Vulnerability to Hydro-Climatic Exposure	62
	4.3.3.1 *Water and Soil Salinity*	*62*
	4.3.3.2 *Tropical Cyclone*	*63*
	4.3.3.3 *Land Erosion*	*65*
4.4	Conclusion	65
	References	66

Experiencing Climate Change in Bangladesh
http://dx.doi.org/10.1016/B978-0-12-803404-0.00004-1

4.1 INTRODUCTION

The purpose of this chapter is to introduce the case study area and the back-drop of socio-ecological and institutional settings in the coastal areas of Bangladesh. After a brief overview of the physical setting of the coastal areas in the next section, this chapter goes on to describe the important socio-economic features of the coastal zone in general and the southwestern region in particular. The chapter further explores how the major hydro-climatic hazards have been affecting the livelihoods, and thereby the well-being, of the households in the coastal zone. The subsequent section analyzes the information on potential future climate variability and changes and their likely damage to the coastal areas of Bangladesh. Much of the analysis of the socio-economic components in general is based on secondary data from the Bangladesh Bureau of Statistics (BBS) on the 19 districts delin-eated as the coastal zone of Bangladesh (Figure 4.1).

Having initially examined the context of the coastal zone as a whole, an exploration of the local level conditions then follows in the second part of the chapter. The second part proceeds as follows. The first section sets out a

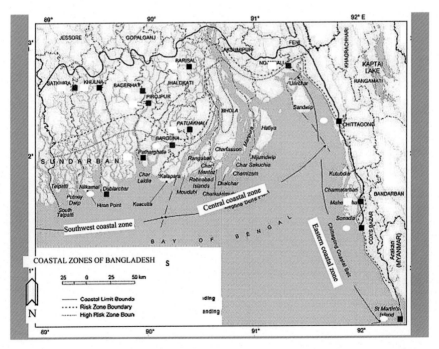

Figure 4.1 Map of the coastal region of Bangladesh. *Source: Adapted from FAO (2014).*

profile of the case study site. It presents the location, physical features, demographics, local infrastructure, public facilities, and existing livelihood activities. This is followed by a discussion on the current vulnerabilities of the households due to hydro-climatic exposure. The final section draws some preliminary conclusions regarding community characteristics and the vulnerability context of the region.

4.2 COASTAL BANGLADESH

4.2.1 Physical Setting

The coastal region of Bangladesh is about 47,201 km², which covers nearly 32%, that is, 2.85 million hectares (Figure 4.1), of the area of the country. At the southern end, the country is connected to the Indian Ocean through a coastline of 710 km along the Bay of Bengal, extending from the tip of Teknaf peninsula in the southeast to the southwestern coast of Shatkhira district.

The majority of the coastal area lies within the delta of the Ganges–Brahmaputra–Meghna (GBM) river system, which is considered to be a transboundary river basin with a total area of approximately 1.7 million km², spread over India (64%), China (18%), Nepal (9%), Bangladesh (7%), and Bhutan (2%) (FAO, 2011). After flowing into Bangladesh, the Ganges, the Brahmaputra, and the Meghna join and drain the GBM basin through the country into the Bay of Bengal.

The coastal region, the lowest landmass of Bangladesh, is characterized by a large network of river systems, dynamic estuaries, interaction between huge freshwaters carried by the river systems, and a saline waterfront infiltrating inward from the seal. The whole coastal area is divided into three coastal regions, the eastern, central, and western regions (Figure 4.1). The eastern coastal region is a narrow strip that extends from the southern tip of the mainland to the Feni river estuary. This part is characterized by small hills running parallel to the coastline and flat beaches. The central coastal area is morphologically very dynamic. This region receives most of the combined flow of the Ganges–Bhrahmaputra–Meghna river system before discharging into the Bay of Bengal. Huge sediment deposits from these rivers and a strong current result in a high rate of erosion and accretion that are observed in this part of the coastal region. The western coastal region is an inactive deltaic area where the Sundarbans, the largest naturally occurring mangrove forest of the world, is located. The landforms of this area include mangrove swamps, tidal flats, natural levees, and tidal creeks.

The western coastal zone lies at 0.8–4 m above the mean sea level (Iftekhar and Islam, 2004).

The coastal regions comprise various ecological systems, including tidal floodplains, brackish water estuaries, mudflats of Sundarbans mangroves, islands, accreted land, beaches, and a peninsula. The coastal flat plain is mainly used for crop production, livestock rearing, salt farming and shrimp farming activities. Coastal areas are endowed with very rich aquatic biodiversity that includes extremely diverse fish fauna, shrimps, crabs, molluscs, and mangroves (Iftekhar, 2006; Ahamed et al., 2012).

Out of 64 districts in Bangladesh, the coastal zone consists of 19 districts that undergo three basic geophysical processes: tidal water fluctuation, salinity intrusion and risk from cyclone and storm surges. In addition to the coastal plains, there are more than 70 small islands in the coastal areas in Bangladesh.

4.2.2 Socio-Economic Context

4.2.2.1 Population

Table 4.1 shows the basic demographic information of the coastal areas in relation to the national characteristics of Bangladesh. The population in the coastal zone, according to the census of 2011, is approximately 40 million, representing 26.73% of the total population of the country. Its average density of 862 persons per square kilometer is below the mean of 1015 people per square kilometer overall for Bangladesh.

In the interior coastal districts, population density is higher than in the exterior coastal areas and remote islands. The population of the coastal zone has grown by 0.75% a year for the past 10 years (2001–2011), markedly below the 1.37% population growth for the country as a whole, indicating a net out-migration from the coastal areas to other parts of the country.

Table 4.1 Basic demographic indicators of the national and coastal zone for the year 2011

Indicators	Coastal zone	National
Population (million)	40.04	149.77
Male (%)	49.40	50.06
Female (%)	50.60	49.94
Density (population/km^2)	862	1015
Intercensal growth (%)	0.75	1.37
Household size	3.9	4.4

Data source: BBS (2012); complied and analyzed by the researcher.

Over 77% of the 40 million coastal people live in rural areas. The literacy rate of the population aged 5 years and above, at the national level, is 55.08% for both sexes, and the corresponding figure in the coastal areas is a little higher, at 56.14%. Overall, the female population has a lower literacy rate than their male counterparts.

4.2.2.2 Poverty and Unemployment

Poverty, unemployment, and malnutrition are prevalent in Bangladesh. The Monitoring of Employment Survey 2009 estimated the total labor force of the country at 53.7 million (15+ years), of which 28.7% were categorized as underemployed (BBS, 2009). At present, approximately 53% of the country's rural population are classified as poor (World Bank, 2013a). Poverty encompasses the notion of human deprivation, in the form of lack of access to basic needs and services, and absolute poverty is defined as less access to the required caloric intake (2122 kcal per day per person). Compared to the rest of the country, income poverty in the coastal zone is higher (WARPO, 2004). Poverty is the immediate underlying cause of widespread malnutrition in Bangladesh, especially among children and women. The rates of malnutrition in Bangladesh are among the highest in the world (FAO, 2013a).

4.2.2.3 Livelihood Characteristics

The bio-physical conditions in the coastal areas mostly dictate the livelihood activities of coastal households. Some of the livelihood activities in this area are very similar to mainstream income activities found everywhere in the country, such as crop agriculture and rearing livestock or poultry, but certain activities are unique to the coastal zone. However, among the coastal districts the prevalence of these specific livelihood activities differs to a great extent. These unique livelihood opportunities can be exclusively attributable to the speciality of the bio-geophysical systems of this area. Some of these livelihood activities are shrimp farming, shrimp fry and crab collection, crab fattening, salt farming, coastal and marine fishing, fish processing (drying), and extraction of forest products, especially from the Sundarbans.

4.2.2.4 Agriculture and Structure of Land Distribution

Bangladesh is predominantly an agricultural country. This sector alone accounts for 43.6% of the employment of the labor force, and its direct contribution to the gross domestic product stood at 14.33% in the fiscal year 2012–2013 (World Bank, 2013a).

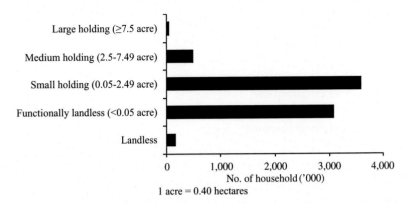

Figure 4.2 Types of coastal household by landholding size. *Data source: BBS (2008); complied and analyzed by the authors.*

Land has become a scarce resource in Bangladesh, and land ownership is highly skewed. Because of demographic pressures and accelerated urbanization, the country's cultivated area has been decreasing at a rate of 1% a year (World Bank, 2013a). According to the agriculture census 2008 of the Bangladesh Bureau of Statistics, nearly 44% of rural households in the coastal zone are functionally landless (households owning <0.05 acre of land), and about 12.66% households are absolutely landless. There are about 3.6 million small farming households with an operating area of less than 2.5 acres; they are the largest livelihood group, comprising 48% of the total households. Figure 4.2 shows the categories of households based on the size of the land that they own.

There are approximately 2.2 million agricultural labourers in the coastal districts, which marks the second largest occupational group, accounting for 30% of coastal rural households (BBS, 2008). In the case of nonfarm livelihood groups, whose principal occupation is not agriculture, fishermen are the single largest group.

4.2.2.5 Shrimp Farming

Bangladesh, with its 710 km of coastline, estuaries of numerous rivers and tributaries, and low-lying tidal floodplains provides enormous opportunities for brackish water shrimp aquaculture (Alam et al., 2005; USAID, 2006). Shrimp is a particularly valuable crop. It contributes to Bangladesh's national economy significantly in the way of foreign income earning. In terms of foreign exchange earnings, the shrimp sector ranks second in Bangladesh, generating a total export value of approximately US$380 million

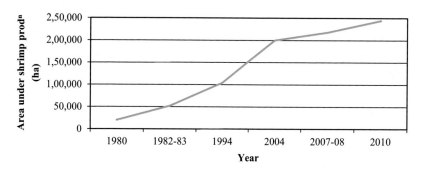

Figure 4.3 Expansion of area under shrimp farming over 30 years. *Data source: USAID (2006), Belton et al. (2011) and FAO (2013b); complied and analyzed by the authors.*

in the 2008–2009 fiscal year (DOF, 2010). The total area under brackish water shrimp cultivation was less than 49,500 acres in 1980; by 2010, the land assigned to shrimp ponds had risen to approximately 605,000 acres (Figure 4.3), representing a 12.25-fold increase (USAID, 2006; Belton et al., 2011; FAO, 2013b). It is estimated that approximately 1.2 million people are directly dependent on shrimp production, of which about 185,000 people are involved in fry collection (USAID, 2006; BFRI, 2011).

Since the early 1970s, brackish water shrimp aquaculture started to appear in the southwest coastal belt, mainly by converting rice fields into shallow earthen ponds (locally known as gher), which are located by the rivers or canals and impounded by the Bangladesh Water Development Board's polders or dykes, which were originally constructed to support agriculture on the coastal flats (Karim, 1986).

4.2.2.6 Fishing

In the coastal areas, fishing is the dominant income source of about 14% of households (WARPO, 2004). They operate in the rivers, estuaries, coastal waters, and intertidal areas such as mangrove areas and sometimes in the deep sea. According to the estimation of the Department of Fisheries (DOF), there are about 1 million people involved in coastal and marine fisheries, operating around 44,000 mechanized and nonmechanized boats, mostly catching shrimp and finfish on the continental shelf (DOF, 2010). With the growing investment in fishing boats and gear by entrepreneurs, as well as the increasing poverty, more and more people are encroaching on traditional fisheries, causing over exploitation of resources that is manifest in declining fish stocks (WARPO, 2004).

4.2.2.7 Institutional Environment

Bangladesh is a democratic republic with a unicameral parliament. The Prime Minister is the chief executive of the country selected by the President from the leaders of the majority party. For the convenience of administration, the country is divided into seven administration divisions, and each division is further subdivided into districts. There are 64 administrative districts and 485 upazilas (subdistricts) in the country. All of the divisions, districts, and Upazila headquarters are urban centers. Below the level of Upazila, there are rural micro areas know as Unions, and each Union consists of multiple villages.

The local government system in Bangladesh consists of single-tiered city councils or municipality and three tiered rural local government, comprising a district council (Zila Parishad), subdistrict council (Upazila Parishad), and union council, and three hill district parishad. Rural and urban local government bodies are entrusted with a range of functions relating to community welfare and local development. However, local government as a political institution has yet to take proper shape with real power, functions, and resources to ensure public participation in development activities.

4.2.2.8 Role of NGOs

In the context in which the level of mistrust in the public sector has been high and the ability of the state to provide sufficient services to a growing population has been in decline, nongovernmental organizations (NGOs) sprang up to respond to the challenges and occupy the center stage in the development landscape of Bangladesh (Devine, 2003; Zohir, 2004). There are 22,000 NGOs currently operating in the country, and they have managed to extend their activities in more than 78% of rural villages in Bangladesh (World Bank, 1996; NGO News, 2013). The NGOs offer a wide range of services that are fundamental to people's day-to-day livelihoods, including micro-credit, education, health and sanitation, family planning, and so forth (Devine, 2003; Zohir, 2004). With the high prevalence of NGOs, the poor people are able to diversify their livelihood options (WARPO, 2004). It is an important social asset for poor people to develop an effective relationship with NGOs in their pursuit of sustainable livelihoods in the wake of risk and vulnerability (Devine, 2003).

4.2.3 Characteristics of the Climate

Bangladesh generally enjoys a subtropical monsoon climate characterized by wide seasonal variations in rainfall, moderately warm temperatures, and high humidity. The climatic subregions of the country are shown in Figure 4.4.

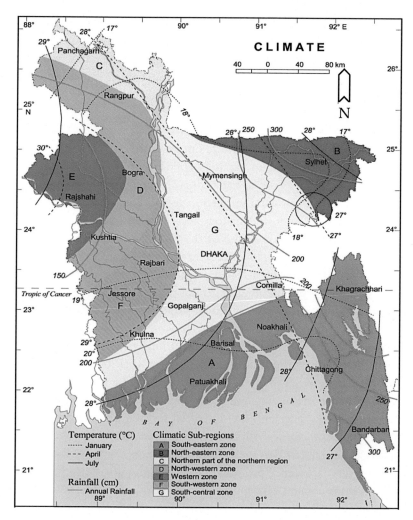

Figure 4.4 Map of the climatic subregions of Bangladesh. *Source: Adapted from Rashid (1991).*

There are four prominent seasons in a year: winter (December–February), premonsoon (March–May), monsoon (June–September), and postmonsoon (October–November). Temperature in the winter fluctuates from a minimum of 7.22–12.77 °C to a maximum of 23.88–31.11 °C. As winter progresses into the premonsoon summer season, temperatures rise and reach the maximum in April. The maximum summer temperature ranges between 30 °C and 40 °C, and occasionally higher in some places. The summer monsoon lasts from June through October. During the monsoon season, Bangladesh receives its maximum rainfall, which makes up 80% of the total annual rainfall and which occasionally causes disastrous flood. Annually, the country receives at least 2000 mm of rainfall. The average annual rainfall in the western region is 1600 mm. The countries' annual rainfall sometime exceeds 4000 mm.

4.2.4 Biophysical Hazards and Key Impacts
4.2.4.1 Waterlogging and Drainage Congestions
Beginning in the 1960s, approximately 6000 km of embankment has been constructed along the coastline, banks of rivers, and coastal estuaries to polder the areas so as to provide protection to the coastal population against flooding from high tides. Regulators and flushing inlets are provided to control water intake and drainage of the rainfall run-off from the polder areas by gravity flow during low tide. Currently there are 139 polders across the coastal zone of Bangladesh. The embankment project enabled people to increase agriculture production but cannot protect their lives and properties against the inundation caused by storm surges, as it was designed only for controlling tidal flood.

Siltation in the peripheral rivers surrounding the embankments and poor maintenance of the drainage channel network of the polders have contributed to internal drainage congestion and heavy external siltation. This has caused waterlogging and water stagnation problems in parts of the coastal areas. These problems are amplified further when there is heavier than usual rain coinciding with the release of water from barrages in India, especially the Farakkah Barrage and the Durgapur/Damodar Barrage. The coastal districts experience prolonged waterlogging situations that result in significant displacement of population, disruption of livelihoods, and serious damage to agriculture and aquaculture. Figure 4.5 depicts waterlogging

Figure 4.5 Waterlogging in coastal area. *Source: ECB Project (2011).*

in one of the southwestern coastal districts as a result of unusual rain in 2011.

Excessive monsoon rain at the end of July and early August in 2011 caused a prolonged waterlogging situation in the southwestern coastal districts of Satkhira, Khulna, and Jessore, where more than 900,000 people were estimated to be affected. Many of the affected people had to temporarily leave their houses and took shelter in community buildings, such as schools, colleges, cyclone centers, or other high ground, and they stayed for 3–4 months in some areas while waiting for the water to recede to return to their homes (ECB Project, 2011).

4.2.4.2 Salinity Intrusion

Water and soil salinity is a common problem in different coastal districts in Bangladesh. It results in significant losses of ecosystem functions and services, such as soil degradation, deforestation, destruction of homestead vegetation, loss of coastal vegetation, and loss of micro flora and fauna (Paul and Vogl, 2011; Sohel and Ullah, 2012). According to the Salinity Survey Report 2010 of Bangladesh, about 2.5 million acres of cultivated land are affected by soil salinity to varying degrees, accounting for 70% of the total cultivable area (SRDI, 2010).

Salinity greatly affects crop production, depending on the level of salinity and the tolerance limit at the critical stage of the growth of a particular crop. It restricts cultivation of rice and dry season rabi crops (agricultural crops sown in winter and harvested in the spring), leading to changes in land use patterns in the coastal regions.

With increasing salinity in surface and groundwater, access to freshwater becomes scarcer, and that, in turn, affects economic and household activities. Surface water salinity in the coastal districts of Patuakhali, Pirojpur, Barguan, Satkhira, Bagerhat, and Khulna has risen by 45% since 1948 (IRIN, 2013).

4.2.4.3 Tropical Cyclone and Storm Surge

Figure 4.6 shows the paths of the historical cyclonic storms of the Bay of Bengal that hit the Bangladeshi coast.

The Bay of Bengal, bordering the Bangladeshi coast, has historically been an area for forming the deadliest cyclones in the world, accounting for about 80–90% of the global loss of lives and property associated with cyclones (Chowdhury, 2002). Cyclones strike the coastal regions of Bangladesh in early summer (April–May) or the late rainy season (October–November).

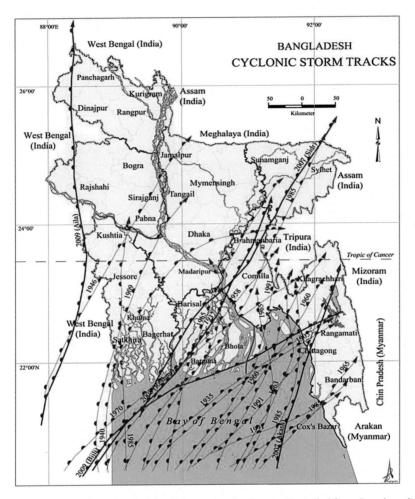

Figure 4.6 Map of historical cyclonic storm tracks. *Source: Compiled from Banglapedia (2013).*

On average, a severe cyclone strikes Bangladesh every 3 years (MOEF, 2009). Fifty percent of all the cyclones that have caused 5000 deaths or more have taken place in Bangladesh (MOFDM, 2008). Indeed, the number of deaths is very much dependent on a single extreme event, as happened in the deadliest cyclones of 1970 and 1991 when approximately 300,000 and 139,000 lives were lost, respectively (Haque et al., 2011). Table 4.2 provides a list of cyclones and human death tolls from these cyclones (BMD, 2007 cited in DDM, 2013).

Table 4.2 Major cyclones that hit the Bangladeshi coast, and death tolls

Year	Maximum wind speed (km/h)	Storm surge height (m)	Death toll
1965	161	3.7–7.6	19,279
1965	217	2.4–3.6	873
1966	139	6.0–6.7	850
1970	224	6.0–10.0	300,000
1985	154	3.0–4.6	11,069
1991	225	6.0–7.6	138,882
1997	232	3.1–4.6	155
2007	223	–	3363
2009	92	–	190

4.2.5 Potential Impacts of Climate Change

Bangladesh has a typical monsoon climate. For the Asian monsoon system, there is a large uncertainty about how this system would respond to climate change, especially the most important climate variable–precipitation induced by monsoon (Seneviratne et al., 2012). The same is true for projections for future tropical cyclones and their related damage in the coastal areas of Bangladesh due to climate change.

However, it is also estimated that currently 8.06 million inhabitants in coastal Bangladesh are vulnerable to storm surge–related inundation to a depth of more than 1 m; the number will increase by 68% with an average 1% population growth by 2050, even without climate change, and by 110% by 2050 in a changing climate in the absence of further adaptation measures (World Bank, 2010).

A number of studies have shown that Bangladesh is one of the most vulnerable counties to the impacts of sea level rise (World Bank, 2013b; Karim and Mimura, 2008). A 10% intensification of the current 1-in-100-year storm surge, coupled with a 1 m SLR, could affect 23% of the total coastal area of Bangladesh (DECC, 2011). Projected elevated sea levels could cause large-scale intrusion of saltwater into the freshwater systems, drainage congestion, and waterlogging in the delta during high flow periods, and could change the dynamics of erosion and accretion of the river and estuary banks (MOEF, 2012).

4.3 DESCRIPTION OF THE STUDY AREA

Mongla is an Upazila (subdistrict) of Bagerhat District located in the south-western coastal region of Bangladesh (Figure 4.7). This sub-district covers an area of 1461.22 km² of which 1083 km² is forest area. For administration

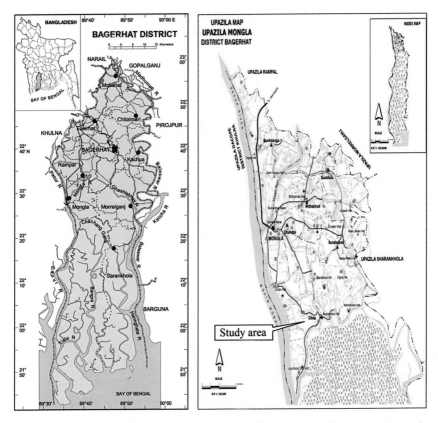

Figure 4.7 Maps showing location of Mongla Upazila in Bagerhat district and the study area. *Source: Compiled from Banglapedia (2013) and LGED (2013).*

purposes, the Upazila is divided into six unions, each of which comprises several villages. The total population of this subdistrict in 2011 was 142,358 according to population and housing census 2011 (BBS, 2012). This subdistrict had a population density of 98 persons per square kilometer in 2011, compared to 1015 persons per square kilometer for Bangladesh as a whole. Mongla and Passur are the two main rivers in this subdistrict. The Bhairab or Rupsha River flows to south of Khulna and is renamed as Passur near Chalna. On the western boundary of the Mongla subdistrict, the Poshur River flows south into the Sundarbans, and, flowing farther south, the river finally discharges into the Bay of Bengal. This river is an important river route for the country, through which local and regional water vessels ply and marine vessels enter the Mongla Port, the second sea port of the country.

4.3.1 Geography of the Study Area

Chila, selected for the fieldwork, is located about 5 km to the south of Mongla Upazila Headquarters, and is accessible via a tarmac road. Chila is bounded by the Poshur River in the west, Chandpi union in the east, and the Chila River in the north (Figure 4.7). The Chandpi range of the world's largest mangrove forest, Sundarbans, is located in the south, and farther south is the Bay of Bengal. The area of this union is about 30 km². For administrative purposes, Chila is divided into 14 small villages.

North of the Chila, a small river also named Chila flows to the west and falls into the Passur River that runs on the western boundary of the union. In addition to these two rivers, there are a number of canals crisscrossing the Chila Union that together constitute common water bodies primarily used for water transport, fishing and shrimp farming. The water flow within this area is mainly regulated by the Passur River. The Passur, and all its distributaries, are tidal channels through which tidal water enters into the study area during high tide and drains out during ebb tide.

4.3.2 Socio-Economic Features

4.3.2.1 Population Dynamics

Table 4.3 shows some basic demographic information about Chila. As per the information provided by the office of the Chila Union Council, the population of Chila was 19,150 in 2012, corresponding to 4794 households. The sex ratio was 104 males per 100 females. The average population density of this area is 635 persons per square kilometer. The age–gender structure of the population in the surveyed households is shown in Figure 4.8. The highest proportion of population is adults, between 25 and 65 years of age, accounting for about half of the total population. This is followed by youth (15–24 years) and children (<15 years).

Table 4.3 Basic demographic indicators of Chila

Indicators	Frequency
Total population	19,150
Number of households	4794
Male (%)	49
Female (%)	51
Density (population/km²)	635
Literacy (%)	60

Source: Chila Union Council (2013).

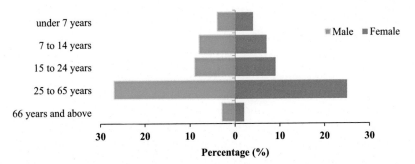

Figure 4.8 Distribution of population by age and gender. *Source: Chila Union Council (2013).*

The population belonging to the age group of 66 years and above constitute only 5% of the total population. This pattern suggests that the historical high-fertility regime of the study area has started to stabilize or decline, which corresponds to the population dynamics of the rest of the country.

4.3.2.2 Health and Education Facilitates

Like other rural parts of the country, the provision of public services in Chila is not well developed. However, people have access to a range of public services, including education, health care, postal, and cyclone shelter, and an orphanage (Table 4.4).

There are 12 primary schools, four secondary schools, and three madrasas (Islamic education) in Chila. The government provides books in the primary school free of cost. Health care services throughout the union are provided by a doctor and a family welfare visitor, assisted by a number of family welfare assistants housed in a Union Welfare Center (UWC). In addition, there

Table 4.4 Public facilities in Chila

Indicators	Frequency
Primary school	12
High school	4
Madrasa	3
Health complex	1
Community clinic	2
Veterinary clinic	1
Orphanage	2
Cyclone center	12

Chila Union Council (2013).

are two community clinics built on community-donated land to provide family planning, preventive, and limited curative services. This union also has a facility for poultry vaccination and animal health treatment.

4.3.2.3 Other Public Services

Housed in the Chila Union Council Office is an information and service center intended to provide government forms, examination results, birth and death registration, livelihood information, computer training, and other services. It has three post offices and two orphanages, and also holds four village markets that attract buyers and sellers from surrounding villages.

4.3.2.4 Road Infrastructure

The Upazila headquarters, on which the lowest tier of the government offices is located, is connected to Chila by an 11-km-long main road, which is the only paved road in this union. Most of the villages within the union, local markets, farms, and ghats (steps leading to riverbank) are connected by an 80 km herringbone and/or earthen road, although many areas are difficult to reach. During the rainy season, the earthen roads become impassable. Rickshaw vans and motor bikes are the only mode of transport for ferrying people and goods.

In addition to road communication, the Passur River along with the Chila River is used as the main water communication system in the study area. Through this river route, people and goods are ferried locally as well as regionally throughout the year.

4.3.2.5 Livelihood Activities

The livelihoods of the study area are centered on shrimp aquaculture, fishing, and agriculture. Currently there are 5381 registered shrimp ponds in Mongla Upazila, covering 26,900 acres of land (e.Mongla, 2013). Other activities include crab fattening, petty trade, migrant workers, wage labor, and collection of fuel wood and honey from the nearby Sundarbans Forest. Table 4.5 shows the distribution of seasonally varying livelihood activities in Chila.

4.3.2.6 Farming

The composition of households in terms of size of the own-account farm is shown in Figure 4.9.

Of the total surveyed household units, about half (46%) are characterized as nonfarm households, of which 84% are functionally landless, operating less

Table 4.5 Season calendar of livelihood activities in Chila

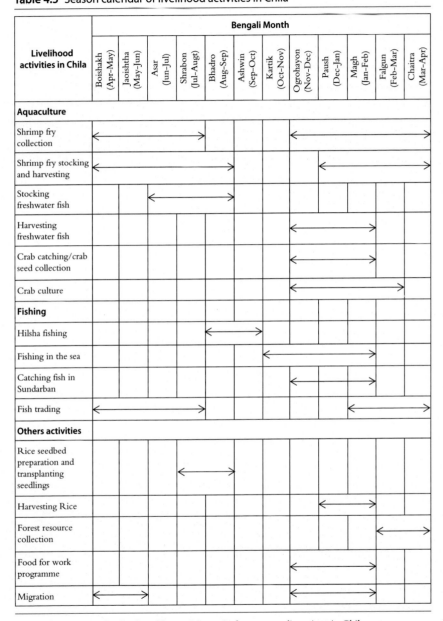

Livelihood activities in Chila	Boishakh (Apr-May)	Jaoishtha (May-Jun)	Asar (Jun-Jul)	Shrabon (Jul-Augt)	Bhadro (Aug-Sep)	Ashwin (Sep-Oct)	Kartik (Oct-Nov)	Ogrohayon (Nov-Dec)	Paush (Dec-Jan)	Magh (Jan-Feb)	Falgun (Feb-Mar)	Chaitra (Mar-Apr)
Aquaculture												
Shrimp fry collection	←	—	—	→				←	—	—	—	→
Shrimp fry stocking and harvesting	←	—	—	—	→			←	—	—	—	→
Stocking freshwater fish			←	—	→							
Harvesting freshwater fish								←	—	→		
Crab catching/crab seed collection								←	—	→		
Crab culture								←	—	—	→	
Fishing												
Hilsha fishing					←	→						
Fishing in the sea							←	—	—	→		
Catching fish in Sundarban								←	—	→		
Fish trading	←	—	—	→				←	—	—	—	→
Others activities												
Rice seedbed preparation and transplanting seedlings				←	→							
Harvesting Rice								←	→			
Forest resource collection											←	→
Food for work programme								←	—	→		
Migration	←	—	→					←	—	→		

Source: Seasonal calendar developed by participants in focus group discussions in Chila.

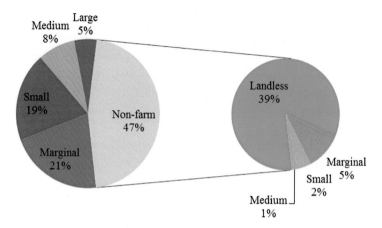

Figure 4.9 Distribution of households by farm type.

than 0.05 acres (0.02 ha) of land, and the rest have rented out their lands on cash or share tenancy arrangements. Farmers of marginal-to-small farms accounted for 40% of the total households whose farm size ranged between 0.05 and 2.49 acres. The corresponding figures for medium (2.50–7.49 acres) and large farm (≥7.5 acres) households were 8% and 5%, respectively.

In the study area, almost all farming households (97% of the survey population), regardless of the farm size, were engaged in gher farming. Although gher farming is often conceived of as brackish water shrimp farming, in practice farmers often undertake multistocking of fries: shrimp, prawn, and freshwater fish, locally known as white fish. In many cases, people have more than one type of activity as their livelihood strategy. According to this field survey, a majority of the surveyed individuals (90%) practice shrimp farming along with other backward- or forward-linking activities, such as fry catching, fry trading, shrimp farming, and shrimp trading.

In Chila, 525 shrimp farms are in operation, covering more than 5000 acres of land. Shrimp aquaculture is being carried out mainly by converting rice fields into shallow earthen ponds (known locally as gher), which are located by rivers or canals. The ghers are impounded by the Bangladesh Water Development Board's polders or dykes, which were originally constructed to support agriculture on coastal flats. Shrimp ghers are connected to the estuaries and canals through channels and sluice gates to exchange brackish water, which is very important for trapping wild shrimp fry, as well as providing a source of natural food and maintaining salinity levels (Paul and Vogl, 2011). Currently, trapping of wild shrimp fry has largely been replaced by stocking of shrimp fry, either produced in hatcheries and/or collected from natural water courses. Shrimp-bred postlarvae (PL) collected in the estuarine nursery areas are highly damaging, and the rate of damage is 1:34 to 1:36, meaning that 34–36 other aquatic fauna are destroyed to collect a single shrimp fry (BFRI, 2011). PL collection, on the other hand, restricts recruitment of the fish to migrate to sea to complete their life cycle. Moreover, large quantities of mud snails are collected to feed the prawn population in the ponds (Belton et al., 2011)

The salinity regimes in a shrimp pond undergo seasonal variations. During high-salinity periods, between January and July, the shrimp and euryhaline fish species (able to live in both freshwater and saltwater) are stocked and harvested; after that, freshwater fish and prawns are cultured together with brackish water shrimp during August and December, when lower salinity levels prevail in the pond environment. In the monoculture system, only shrimp are cultured year-round, leaving no room for crop production. The current shrimp aquaculture system is extensive and traditional, with little or no management. As a result, returns are also low and variable, ranging from US$75–860 per hectare per year (BDT2000–2300 per acre) (USAID, 2006).

4.3.2.7 Fisheries

The fishery resources in the study area are rich in diversity, with mainly brackish water to minor freshwater fish habitats. The network of rivers, canals, tidal creeks, and tidal and nontidal floodplains of this region provide support for a number of marine and freshwater fishes. Therefore, the livelihoods of the study area are centered on the fishery sectors, which can be classified into four groups: (1) inland capture (open water); (2) inland culture (closed water); (3) marine commercial or trawl fishing; and (4) marine artisanal or small-scale fishing. The land use in Chila is dominated by shrimp gher (ponds), which is the most commercially important culture fishery; culture fishery was found to be practiced in homestead fish ponds and a few semiclosed canals. Further discussion on shrimp farming is in the following section. Along with shrimp farming, the natural harvesting and cultivation of crabs was observed as a growing practice in the study area, as in other southwestern parts of Bangladesh. Seed crabs were collected from intertidal creeks, canals, mangrove forests, and rivers, and are stocked in the ponds for 4–5 months to achieve the desirable size before harvesting and selling to market.

Capture fishery comes next after shrimp aquaculture, on which traditional and artisanal fishers depend for their livelihoods. They operate in the brackish to freshwater rivers, canals, creeks, beels (oxbow lakes), offshore waters, and sometimes in the deep sea. Based on the time that they spend on fishing, fishermen in Chila can be classified as occasional, full-time, or part-time. The monsoon months are the main fishing season, and are characterized by inclement weather. Local fishermen reported the dwindling diversity of fishes, as well as a reduced catch per effort, attributing this to multiple factors, including the following: overfishing, obstruction of fish migration routes, changes in the geo-morphological processes of the river and its connectivity, and reduction of spawning and feed grounds because of rapid expansion of culture fisheries and other physical infrastructures in the tidal and nontidal floodplains.

Hilsa (*Tenualosa ilisha*) is the most important single species of fish, accounting for nearly half of the marine catch in Bangladesh (BFRI, 2011). Hilsa is a migratory species, and adults lay eggs in the freshwater in the delta tributaries, which flow into the Bay of Bengal. After being born, juveniles migrate to the Bay of Bengl to mature in the saltwater for 1 or 2 years, before swimming back upstream to spawn. The Hilsa are captured during their migration to the sea as juveniles and when they travel back upstream as adults. Although the study region is not a major spawning ground, Hilsa exist in the Passur River and the adjacent marine and coastal waters, and are harvested between August and October.

4.3.2.8 Agriculture

In relatively upland areas, where soil and water salinity are low, slightly salt-tolerant rain-fed rice is cultivated during the months of July–September. During this time, transplanted Aman rice, either alone or together with freshwater fish and shrimp, is produced. According to the information provided by the agricultural extension office in Mongal Upazila, the total cultivable area of this subdistrict is about 32,170 acres, of which about 97% of the land was used for agriculture during 2009–2010. Almost all of the agricultural production come from rice, mostly transplanted Aman; nonrice crops including winter and summer vegetables, potatoes, and chillies. In Chila Union, only 15% of the surveyed households were found to practice agricultural production along with other livelihood activities. Like other parts of this subdistrict, the agricultural production in the study area is rapidly dwindling because all the cultivable lands are affected by different levels of salinity, which restricts saline-sensitive crop production.

4.3.2.9 Extraction of Sundarbans Forest Resources

Like other surrounding areas of the Sundarbans, many households in Chila depend on the mangrove forest's resources for their livelihood. The livelihood groups who are contingent upon products of the Sundarbans include shrimp fry collectors, fishers, bawalies (wood cutters), mawalies (honey collectors), boatmen, golpata (Nypa palm) collectors, and medicinal plant collectors. People working in the forest encounter dangerous surroundings, including falling victim to Bengal tigers, robbery, kidnapping, and extortion.

4.3.2.10 Seasonal Migration

Short-term seasonal migration as a livelihood option is pursued mainly by male members of the households. Male out-migration mostly takes place from April to May, which is the harvesting season of Boro rice, and from November to February. As Boro harvesting season creates a seasonal peak in the labor requirement for agriculture, poor households send their male members to neighbouring rice-growing districts for a couple of weeks, where the wage rate is relatively higher than in the local labor market. The months from November until February year were most frequently mentioned as the months when male migration appeared as a characteristic of poor households' livelihood strategies because of a dearth of local employment opportunities during that time. In addition to short-term migration, long-term migration to Middle Eastern countries is not uncommon, although the incidence is less than in other parts of the country. Aside from male out-migration, participants in the focus group discussions reported a small migration of poor women from the study area to urban towns and cities to take up placements in domestic services.

4.3.3 Current Vulnerability to Hydro-Climatic Exposure

4.3.3.1 Water and Soil Salinity

All of the rivers in the coastal region are tidal, with semidiurnal, fortnightly, and seasonal variations in water levels fluctuating from 2 to 4.5 m. During every high tide, saline water from the Bay of Bengal enters the estuaries and mixes with the freshwater inflow, and this saline–freshwater front travels upstream with decreasing concentration. Interlinked with the Passur River, being the largest river in this region, the salinity front has encroached inland and penetrated into domestic ponds, groundwater, and agricultural land through various small rivers, canals, and water inlets. The level of water salinity is highly seasonal, and it varies inversely with the variation of freshwater flow from upstream, which depends on rainfall and the release of water from barrages in India.

The Fisheries Office in Mongla regularly monitors the water quality of the Passur and Mongla Rivers with a measurement of salinity during

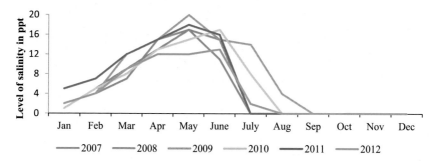

Figure 4.10 Seasonal changes in salinity in the Mongla River. *Data Source: Upazila Fisheries Office, Mongla (2013).*

spring tide and neap tide each month. According to the salinity data for 2007–2012 from the Upazila Fisheries Office, the level of salinity in the Passur River increases gradually from January and reaches a peak in the premonsoon season (Figure 4.10).

With the start of the monsoon rainfall, salinity in the river water appears to decline rapidly and remains below 1 part per trillion (ppt) throughout the postmonsoon season. Because of flooding with saline water from the river and/or seepage of river water, salt ingresses the soil and becomes concentrated in the surface layer as water evaporates from the surface soil. Along with the foregoing factors, rapid expansion of shrimp cultivation in this area plays an important role in salinity intrusion, in which agricultural land remains inundated with brackish water almost all year.

4.3.3.2 Tropical Cyclone

Chila suffers considerably from the impacts of frequent tropical cyclones and storm surges. In recent times, it was battered by two severe cyclones like many other areas in the southwestern region. In 2007, the area was hit by a supercyclone causing severe impacts in terms of losses of human life and material damage, and the government identified the area as the "worst affected" union. A severe cyclone named "Aila" struck this area again in 2009, with a storm surge of up to 6.5 m. In this instance, there was no death toll, but the damage to properties and public infrastructure was significant across the region. The information collected from the Chila Union Council indicates that almost all of the households in this union were affected to varying degrees, along with damage to public infrastructures. Table 4.6 summarizes information from the Chila Union Council on damages and losses caused by cyclone Aila in 2009.

Table 4.6 Information on the damage in Chila caused by cyclone Aila

Sl. No.	Item	Frequency
1	No. of households affected	4500
2	Death toll	0
3	No. of injured persons	43
4	Crop damaged (Bangladeshi Taka)	2 million (US $29.50 thousand)
5	Dead livestock and poultry	664
6	No. of partially damaged educational institutes	15
7	Fully and partially damaged roads (km)	62
8	Fully and partially damaged embankments (km)	5
9	No. of damaged trees	15,000

Cyclone Aila resulted in a scarcity of safe water for drinking, washing, and bathing because of the contamination of safe water sources by saline water inundation, dead animals, and debris. People in this area mostly rely on rainwater harvesting and freshwater ponds or water bodies for drinking and other household purposes. The cyclone also hampered the sanitation facilities significantly, as physical damage to household latrines were fairly common across the Chila Union.

The Chila Union Council report shows that all the 525 shrimp ponds of this union, covering about 5000 acres of land, were severely affected, mostly in two ways: huge amounts of shrimp, fingerlings, and other sweet–water fishes washed away because of pond inundation; and physical damage to ponds resulted from falling trees and branches caused by strong winds. In addition, extensive damage and losses occurred to fishing equipment, such as fishing nets and motorized and nonmotorized fishing boats as a result of this severe cyclone. The losses in shrimp aquaculture inflicted by cyclone Aila was estimated to be BDT 300 million (US $4.43 million), and the damage to fishing equipment amounted to about BDT 9 million (US $0.13 million). Table 4.7 shows the estimated losses in the fisheries sector incurred by the inhabitants of Chila as a result of Aila.

Table 4.7 Information on damages to fisheries sector and its estimated costs

Sl. No.	Item	Number	Estimated loss (BDT)
1	Shrimp ponds	525	300 million (US $4.43 million)
2	Fishing nets	2000	2 million (US $29.5 thousand)
3	Nonmotorized boats	200	2 million (US $29.5 thousand)
4	Motorized boats	100	5 million (US $73.8 thousand)

4.3.3.3 Land Erosion

The coastline of Bangladesh changes its position because of land erosion and accretion resulting from the interaction between the two strong agents of delta-building activities, namel, fluvial and marine processes. The impact of change may be as little as a few meters to more than several kilometers. Major erosion takes place in the Meghna estuary, where, between 1973 and 2000, more than 212,500 acres of land were found to be eroded (BWDB, 2001). Although coastline in the southwestern coastal area is generally non-deltaic and stable and therefore erosion is not a major threat, land erosion along the riverbank and on the foreshore appears to continue, with severe consequences to the livelihoods of those living in tidal floodplains. Land erosion causes the loss of important infrastructure, highly productive agricultural land, and settlements. As a consequence, many large and medium farmers become marginal, and go from marginal to landless and even homeless in many instances. This is one of the most disadvantaged groups in the coastal areas. The earthen embankments constructed along the coastlines, riverbanks, and coastal estuaries to form polders is subject to continuous problems of erosion mostly because of rain, wave action, and the overtopping of storm surges. This breaching of the embankment systems exposes the interior land to the threat of tidal surge and salt water intrusion (WARPO, 2004).

4.4 CONCLUSION

The physical and socio-ecological systems in the coastal region, in many ways, differ from the rest of the country. The physical geography of the region is very low lying, very flat, and dynamic. Under the influence of tidal waters and sweet water supplies from upstream, unique brackish water ecosystems have developed in the coastal areas (Mondal et al., 2013). By taking advantage of the unique bio-physical conditions, households in the coastal areas have developed a range of livelihoods—some similar to those in other parts of the country, such as agriculture, for example, and some, such as shrimp aquaculture and salt framing, unique to coastal areas.

The socio-economic conditions of the coastal areas of Bangladesh are characterized by high population density and, although lower than the national average, widespread poverty and predominantly rural settlement, where the majority of the households for their livelihoods depend mostly on natural resource–based activities.

Along with economic and social risk factors, vulnerability to hydro-climatic hazards, such as tropical cyclones, tidal storm surges, soil erosion, and salinity intrusion are the important dimensions of the livelihood systems in the coastal area. The frequent tropical cyclones, with high tidal surges, cause extensive physical destruction, casualties, damage to crops and livestock, and flooding shrimp farm and low-lying areas. In the study area, settlements and physical infrastructures in proximity to riverbanks, foreshores, and estuaries are risk for land erosion, making the households a disadvantaged group in the area.

This is a starting point to understanding the key features of rural livelihoods in the coastal area that provide households with the capacity to adapt their livelihoods in the face of climate variability and change.

REFERENCES

Ahamed, F., Hossain, M.Y., Fulanda, B., Ahmed, Z.F., Ohtomi, J., 2012. Indiscriminate exploitation of wild prawn postlarvae in the coastal region of Bangladesh: a threat to fisheries resources, community livelihoods and biodiversity. Ocean Coastal Manage. 66, 56–62.

Alam, S.M.N., Lin, C.K., Yakupitiyage, A., Demaine, H., Phillips, M.J., 2005. Compliance of Bangladesh shrimp culture with FAO code of conduct for responsible fisheries: a development challenge. Ocean Coastal Manage. 48, 177–188.

Banglapedia, 2013. http://www.banglapedia.org/HT/B_0032.HTM (accessed 10.06.13).

BBS, 2009. Report on Monitoring of Employment Survey-2009. Bangladesh Bureau of Statistics, Bangladesh.

BBS, 2008. Agriculture Census 2008. Bangladesh Bureau of Statistics, Bangladesh.

Belton, B., Karim, M., Thilster, S., Jahan, K.M., Collis, W., Phillips, M., 2011. Review of Aquaculture & Fish Consumption in Bangladesh. Studies and Reviews 2011-53. The WorldFish Center.

BFRI, 2011. Sustainable Management of Fisheries Resources of the Bay of Bengal. Bangladesh Fisheries Research Institute.

BBS, 2012. Population and Housing Census 2011. Bangladesh Bureau of Statistics, Bangladesh.

BWDB, 2001. Hydro-Morphological Dynamics of the Meghna Estuary, Meghna Estuary Study Project. Bangladesh Water Development Board.

Chila Union Council, 2013. Annual Report.

Chowdhury, K.M.M.H., 2002. Cyclone preparedness and management in Bangladesh. In: BPATC (Ed.), Improvement of Early Warning System and Responses in Bangladesh towards Total Disaster Risk Management Approach. BPATC, Savar, Dhaka, pp. 115–119.

DOF, 2010. Fisheries Statistical Year Book of Bangladesh 2008–2009, vol. 26, No.1. Fisheries Resource Survey System, Department of Fisheries, Ministry of Fisheries and Livestock.

Devine, J., 2003. The paradox of sustainability: reflections on NGOs in Bangladesh. Ann. Am. Acad. Polit. Soc. Sci. 590, 227.

DDM, 2013. Cyclone and Storm Surges. Department of Disaster Management, Bangladesh. http://www.ddm.gov.bd/cyclone.php (accessed 13.06.13).

DECC, 2011. Climate: Projects, Observations and Impacts. Department of Energy and Climate Change, UK.

FAO, 2011. Ganges-Brahmaputra-Meghna river basin. Food and Agriculture Organization of the United Nations. http://www.fao.org/nr/water/aquastat/basins/gbm/index.Stm (accessed 12.08.13).

FAO, 2013a. Nutrition Country Profile: Bangladesh, Food and Agriculture Organization of the United Nations. http://www.fao.org/ag/agn/nutrition/bgd_en.stm (accessed 10.06.13).

FAO, 2013b. National Aquaculture Sector Overview: Bangladesh. Fisheries and Aquaculture Department. Food and Agriculture Organization of the United Nations. http://www.fao.org/fishery/countrysector/nasobangladesh/en (accessed 05.03.13).

FAO, 2014. Fisher and Aquaculture Country Profile: The People's Republic of Banladesh. Food and Agriculture Organization of the United Nations. http://www.fao.org/fishery/facp/BGD/en (accessed 02.09.14).

Haque, U., Hashizume, M., Kolivras, K.N., Overgaard, J.O., Das, B., Yamamoto, T., 2011. Reduced death rates from cyclones in Bangladesh: what more needs to be done? Bull. World Health Organ. 90, 150–156. http://dx.doi.org/10.2471/BLT.11.088302.

Iftekhar, M.S., Islam, M.R., 2004. Degeneration of Bangladesh sundarbans mangroves: a management issue. Int. For. Rev. 6 (2), 123–135.

Iftekhar, M.S., 2006. Conservation and management of the Bangladesh coastal ecosystem: overview of an integrated approach. Nat. Resour. Forum. 30, 230–237.

IRIN, 2013. Bangladesh: Rising Sea Levels Threaten Agriculture. Integrated Regional Information Networks. http://www.irinnews.org/report/75094/bangladesh-rising-sea-levels-threaten-agriculture (accessed 13.06.13).

Karim, M., 1986. Brackish Water Aquaculture in Bangladesh: A Review. Ministry of Fisheries and Livestock, Bangladesh.

Karim, F.M., Mimura, N., 2008. Impacts of climate change and sea-level rise on cyclonic storm surge floods in Bangladesh. Global Environ. Change 18 (3), 490–500.

LGED, 2013. Upazila Map: Upazila Mongla. http://www.lged.gov.bd/UploadedDocument/Map/KHULNA/bagerhat/mongla/mongla.jpg (accessed 14.05.13).

MOEF, 2009. Bangladesh Climate Change Strategy and Action Plan 2009. Ministry of Environment and Forests, Bangladesh.

MOFDM, 2008. Cyclone Sidr in Bangladesh: Damage, Loss and Needs Assessment for Disaster Recovery and Reconstruction. Ministry of Food and Disaster Management, Bangladesh, Dhaka.

MOEF, 2012. Second National Communication of Bangladeshi to the United Nations Framework Convention on Climate Change. Ministry of Environment and Forest, Bangladesh.

Mongla Upazilla Bagerhat, 2013. http://monglaupazilla.com/2010-03-20-14-57-06 (accessed 05.03.13).

Mondal, M.S., Jalal, M.R., Khan, M.S.A., Kumar, U., Rahman, R., Huq, H., 2013. Hydrometeorological trends in southwest coastal Bangladesh: perspective of climate change and human interventions. Am. J. Clim. Change 2, 62–70.

NGO News, 2013. NGO List of Bangladesh. http://ngonewsbd.com/ngo-list-of-Bangladesh (accessed 14.06.13).

ECB Project, 2011. Flooding and Prolonged Water Logging in South West Bangladesh: Coordinated Report. http://foodsecuritycluster.net/sites/default/files/FSC_BAN_Flooding_%20%26_Prolonged_Water-logging%20South_West_%20Bangladesh_Coordinated_Assessment_ Report_ Sep11.pdf (accessed 10.06.13).

Paul, B.G., Vogl, C.R., 2011. Impacts of shrimp farming in Bangladesh: challenges and alternatives. Ocean Coastal Manage. 54, 201–211.

Rashid, H., 1991. Geography of Bangladesh. University Press Limited, Dhaka.

Sohel, M.S.I., Ullah, 2012. Ecohydrology: a framework for overcoming the environmental impacts of shrimp aquaculture on the coastal zone of Bangladesh. Ocean Coastal Manage. 63, 67–78.

SRDI, 2010. Saline Soils of Bangladesh. Soil Resource Development Institute, Bangladesh.
Seneviratne, S.I., Nicholls, N., Easterling, D., Goodess, C.M., Kanae, S., Kossin, J., Luo, Y., Marengo, J., McInnes, K., Rahimi, M., Reichstein, M., Sorteberg, A., Vera, C., Zhang, X., 2012. Changes in climate extremes and their impacts on the natural physical environment. In: Field, C.B., Barros, V., Stocker, T.F., Qin, D., Dokken, D.J., Ebi, K.L., Mastrandrea, M.D., Mach, K.J., Plattner, G.K., Allen, S.K., Tignor, M., Midgley, P.M. (Eds.), Managing the Risks of Extreme Events and Disasters to Advance Climate Change Adaptation. A Special Report of Working Groups I and II of the Intergovernmental Panel on Climate Change, pp. 109–230.
USAID, 2006. A Pro-Poor Analysis of the Shrimp Sector in Bangladesh. Greater Access to Trade Expansion (GATE) Project. US Agency for International Development.
World Bank, 2013a. Agriculture in South Asia Bangladesh: Priorities for Agriculture and Rural Development. http://web.worldbank.org/WBSITE/EXTERNAL/COUNTRIES/SOUTHASIAEXT/EXTSAREGTOPAGRI/0,contentMDK:20273763~menuPK:548213~pagePK:34004173~piPK:34003707~theSitePK:452766,00.html (accessed 09.06.13).
WARPO, 2004. Living in the Coast: People and Livelihoods, PDO-ICZMP. Water Resources Planning Organization, Bangladesh.
World Bank, 1996. Pursuing Common Goals: Strengthening Relations between Government and Development NGOs. University Press, Dhaka.
World Bank, 2010. Vulnerability of Bangladesh to Cyclones in a Changing Climate. World Bank Policy Research Working Paper 5280.
World Bank, 2013b. Turn Down the Heat: Climate Extremes, Regional Impacts, and the Case for Resilience. World Bank, Washington, DC.
Zohir, S., 2004. NGO sector in Bangladesh: an overview. Econ. Polit. Wkly. 39 (36), 4109–4113.

CHAPTER 5

Household Assets and Capabilities

Contents

5.1 Introduction		69
5.2 Livelihood Capitals		70
5.2.1 Human Capital		70
	5.2.1.1 Household Size	70
	5.2.1.2 Education	71
	5.2.1.3 Health	72
	5.2.1.4 Technical Training	72
5.2.2 Physical Capital		73
	5.2.2.1 Housing	73
	5.2.2.2 Access to Safe Water	73
	5.2.2.3 Production Equipment	74
	5.2.2.4 Cyclone Shelters	75
5.2.3 Natural Capital		75
	5.2.3.1 Land	75
	5.2.3.2 Livestock	76
5.2.4 Financial Capital		78
	5.2.4.1 Income	78
	5.2.4.2 Loan	80
5.2.5 Social Capital		81
	5.2.5.1 Participation in Community Organization	81
	5.2.5.2 Contact with NGOs	82
5.3 Discussion and Conclusion		82
5.3.1 Human Capital		83
5.3.2 Physical Capital		83
5.3.3 Natural Capital		84
5.3.4 Financial Capital		85
5.3.5 Social Capital		85
References		86

5.1 INTRODUCTION

In this chapter, we move from the socio-economic and vulnerability contexts of rural livelihoods in the coastal area described in Chapter 4, to five types of capital assets accessed by the rural households. Within the

Experiencing Climate Change in Bangladesh
http://dx.doi.org/10.1016/B978-0-12-803404-0.00005-3
69

framework of the livelihoods approach, this chapter examines the levels of key assets owned by the households that give them the capability to respond to the impacts of climate change. Having identified a number of variables representing the five capitals of livelihoods influencing the adaptive capacities of households (Table 3.1), this part of the study presents the analysis of those categories of capital at a household level across the community.

5.2 LIVELIHOOD CAPITALS

The household survey contained a number of questions regarding the five categories of capital (human, social, natural, physical and financial) as being indicative of the adaptive capacity of the households to cope with, and adapt to, climate variability, extreme events, and change. A detailed description of these capital assets owned or accessed by the survey households in the study area is provided below:

5.2.1 Human Capital
5.2.1.1 Household Size
The average household size in Chila is four persons per household, which is slightly smaller than the overall rural household size in Bangladesh (4.4 persons/household). The average dependency ratio varies across income groups (Figure 5.1). The survey data suggest that wealthier households tend

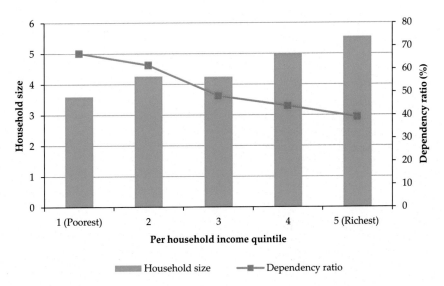

Figure 5.1 Household size and dependency ratio by income group.

to be larger than poorer households, in which the average household size rose from 3.6 for the first quintile group (poorest) to 5.5 for the fifth quintile group (wealthiest). On the other hand, wealthier households have relatively more people of working age than their poorer counterparts, and this phenomenon is measured by a dependency ratio—the ratio of dependents (people younger than 15 years or older than 65 years) to the working age population (aged 16–64 years) and is normally expressed as a percentage.

5.2.1.2 Education

A person who can read and write a sentence in Bengali is assumed to be literate. Table 5.1 shows the literacy rates among different age groups in the survey households.

The literacy rates among the surveyed population aged 5 years and above was 81.3%, which is astonishingly higher than the national average of 55.06%. Literacy varies with gender; 85.7% of men are literate compared with 76% of women. Table 5.1 shows that the percentage of households with young children who do not send their children to school is very low. This has changed since the last decade, as nongovernmental organizations (NGOs), along with the government, have opened schools in many villages so that young children have a chance to obtain an education.

Contrary to other parts of the country, youth and adult literacy across the surveyed population was very high. This is due to the fact that education is highly valued in the study area, as has been observed by a World Bank study in the coastal area of Bangladesh (World Bank, 2010). Regarding educational attainment, in terms of years of schooling, the average number of years of study by male youths and female youths were 8.6 and 8.0, respectively. The corresponding data for adult males and females were 8.0 and 7.0, respectively. Although the literacy level is very high, a higher incidence of poverty caused many children to leave their schools to work with their parents and to help feed their families.

Table 5.1 Rates of literacy in surveyed household members

	Male literacy (%)			Female literacy (%)		
Age group	Literate	Year studied	Illiterate	Literate	Year studied	Illiterate
Child (7–14 y)	98.5	–	1.5	96	–	4
Youth (15–25 y)	96	8.6	4	95.6	8	4.4
Adult (25–65 y)	78.2	8	21.8	67.5	7	32.5
Elderly (>65 y)	60	–	40	20	–	80

5.2.1.3 Health

Figure 5.2 shows the proportion of surveyed respondents in each of the four categories on their self-perceived health status.

The measurement of self-perceived health is subjective, and is expected to capture different dimensions of health that include physical, social, emotional function and biomedical signs and symptoms (Baert and de Norre, 2009). Of the total respondents, some 9% reported their health as poor, 75% as fair, 16% as good, and 0.3% as very good.

5.2.1.4 Technical Training

Skill development and vocational education and training contribute to empowering households for better livelihoods. Related to livelihood skill development training, around 75% of the surveyed people do not have any training whatsoever. Among the training recipients, half of them (52%) had received training on disaster risk reduction provided by the NGOs and government organizations, since the study area is a natural disaster–prone area (Figure 5.3). This was followed by training programs related to shrimp

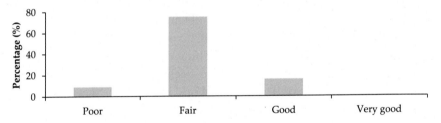

Figure 5.2 Self-assessed health status.

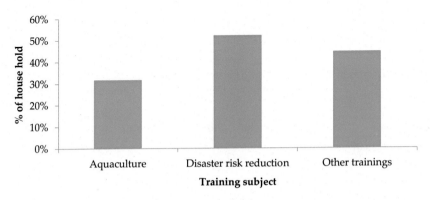

Figure 5.3 Percentage of individuals received technical training.

aquaculture, accounting for 32%. Other training activities included vegetable gardening, crab cultivation, livestock rearing, and some awareness training on climate change and health.

5.2.2 Physical Capital

5.2.2.1 Housing

The economic and social conditions of households that include income, social status, and household size determines the dwelling type in rural Bangladesh. Figure 5.4 shows that most of the people in Chila live in small kacha houses (93% of the survey population), which are made of locally available materials. This type of house is built on a mud floor with the walls made of straw, jute stick, bamboo, mud, or corrugated iron (CI) sheets, with thatch or CI sheets used as roofing. Cement floors, brick-built walls, and concrete roofing (pucca house) are very rare (2%) in the study area. About 5% of the dwelling structures were found to be made with brick-built walls and CI sheet roofing (semi-pucca house).

The survey data indicate that most of the dwelling units in Chilla are poorly built, which cannot provide adequate protection against strong winds, rain, and tidal floods.

5.2.2.2 Access to Safe Water

Like other parts of the southwest coastal region, Chila has a serious problem with drinking water as the natural sources of drinking water have become contaminated due to saltwater intrusion. The levels of salinity clearly show a seasonal pattern, which decreases with the increase of monsoon rainfall and water flow in the downstream network of rivers. During

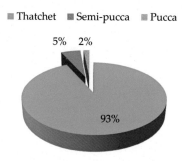

Figure 5.4 House types of the surveyed households.

Table 5.2 Sources of water used by survey population

Source of water	Respondents Percentage of n (number of sample) n = 372
Pond	88 (326)
Rain	99 (369)
River	17 (62)
Purchased water	55 (205)

the dry period, shallow ground water aquifers and numerous rain-fed ponds develop increasing levels of salinity, in addition to water courses. Table 5.2 shows the sources of water used by the sampled households for drinking, bathing, washing, and cooking purposes.

The survey results indicate that people mostly use pond and rain water for drinking and other household purposes, and that they purchase potable water only when the rain water is exhausted and salinity intrudes into ponds.

5.2.2.3 Production Equipment

The status of the ownership of equipment of the surveyed households is presented in Table 5.3.

Among the physical capital assets belonging to households, fishing nets were found to be the most common, followed by fishing boats, because shrimp aquaculture and fishing are the key sources of income and employment in the study area. Agricultural practices have mostly been replaced by shrimp farming, and thus the possession of agricultural equipment is very low in the study area.

Table 5.3 Agricultural and nonagricultural equipment owned by survey population

Equipment type	Respondents Percentage of n (number of sample) n = 372
Fishing nets	36 (135)
Boats	21 (78)
Ploughs	2 (9)
Sewing machines	3 (10)
Rickshaw vans	8 (30)
Motorcycles (used for ferrying people)	1 (5)

5.2.2.4 Cyclone Shelters

Twelve cyclone shelters have been constructed in Chila, which has been hard hit by repeated cyclones. Tropical cyclones, accompanied by storm surges and winds, sweeps away most of the insubstantially built houses in the rural coastal area; thus people take refuge in concrete cyclone shelters. These shelters contributed, to a great extent, to reduce the death tolls in previous devastating cyclones. The salient features of these cyclone shelters are that they are multistoreyed buildings constructed on a pillar basement. The first floor is kept open to allow flood water to flow freely, and people take shelter on the rest of the floors. Now many of these shelters are also being used as schools, especially the ones with multiple rooms.

5.2.3 Natural Capital

5.2.3.1 Land

Land is the most important factor for aquaculture and agricultural production, the primary occupation of the southwest coastal districts. The ownership of land has a strong, positive relationship with income. However, about 39% of households in the study area are landless, meaning that they do not own any cultivable land. The incidence of landlessness for the entire rural area of Bangladesh is 57%. Among those who own cultivable land, the bottom 25% of all households own less than 3.5% of the total land, and at the other extreme, top 10% of households own about 50% of the total cultivated land in the study area. The Lorenz curve in Figure 5.5 shows the land distribution in Chila Union.

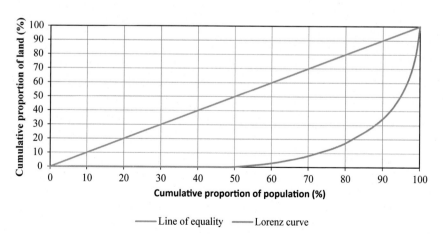

Figure 5.5 Lorenz curve of land distribution in the study area.

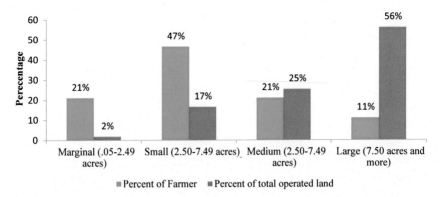

Figure 5.6 Distribution of operated land by farm size in the study area.

The 45° line represents perfect equality in which every household owns same amount of land, and the area below the 45° line and Lorenz curve gives a measure of the extent of inequality. Figure 5.6 demonstrates the distribution of operated land by each of the four farm size groups.

In the study area, about half of the surveyed farmers are small-farm farmers who operate only 17% of the total operable land, whereas the large-farm farmers, comprising only 11% of the total farming households, operate 56% of the total operable land in the study area. The average size of the operated land of farm households in the study area is 2.5 acres, ranging from 0.06 to 26 acres.

5.2.3.2 Livestock

Crop production and livestock farming are closely linked in the Bangladeshi farming system. Cattle, goats, and poultry are the most commonly reared animals in rural Bangladesh. In rural livelihood strategies, livestock rearing is considered as one of the few opportunities for saving, investment, and as a hedge against risk. The study region has been historically characterized by a mixed system of crop production and home-based livestock; however, rapid expansion of shrimp aquaculture in the crop fields, in conjunction with salinity intrusion, has resulted in a reduction in family poultry, goat grazing, and cattle herding.

Family poultry range freely in the household compound and find much of their own food from the surrounding environment, with only a small amount of supplemental feed from householders. Poultry contribute to households' income generation through the sale of birds and eggs, and they are a source of protein in the diet when consumed occasionally. Figure 5.7 shows the poultry flock size of the surveyed households.

In Chila, more than one-fifth of the households do not keep chickens, approximately three-quarters of them keep between 1 and 30 chickens per household, and the remaining 2% of households run small-scale commercial poultry farms, where the flock size ranges from more than 30 to a maximum of 350 birds.

Figure 5.8 shows that most of the households in Chila have given up one of the important income-generating activities, cattle rising. Nearly three-quarters of the surveyed households are unable to keep any ruminants, and the rest maintain a small sized herd, ranging between one and six cattle and between 1 and 10 goats. Only 2% sampled households were found to be able to afford medium-sized cattle farms (7–15 cattle) and the corresponding figure of 1% for goat farms. The main reason the respondents reported, as seen in Figure 5.8, was the lack of feed for cattle and goats in

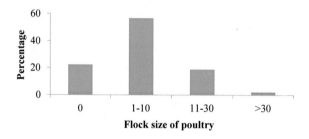

Figure 5.7 Percentage of households by flock size of poultry.

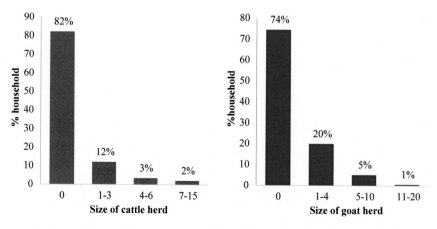

Figure 5.8 Percentage of households by cattle (left) and goat (right) herd size.

the area. Cattle are mostly fed agricultural by-products, such as straw. Goat and cattle are grazed on natural pastures of non-arable land, as well as public wasteland around canals, rivers, roadsides, and railways. Shrub and tree leaves are also used as their fodder, which has traditionally been available in villages. With the increase in soil salinity, however, crop production has been replaced by brackish water shrimp cultivation, and, with the loss of grazing land, the traditional home-based livestock farming has been thwarted.

5.2.4 Financial Capital

5.2.4.1 Income

The economy in the study area was predominantly aquaculture based, with about 60% of households relying on shrimp farming for their livelihoods. The survey data confirmed that gher farming, popularly known as shrimp farming, was the most important source of income for the communities, providing 32% of the total income, as shown in Figure 5.9.

Income from agriculture was probably an important income source before its decline, and, at the time of the survey, made up only 2% of income. Given the importance of farm income, access to land was an important determinant of welfare, and larger farm-holdings were associated with the highest income, as illustrated in Figure 5.10. As shown in Figure 5.10, households were divided into five equal groups, and are ranked from lowest to highest according to total income.

Figure 5.11 shows the Lorenz curve of income distribution among the surveyed households. The Lorenz curve plots the cumulative percentage of total income on the cumulative percentage of households. It was observed

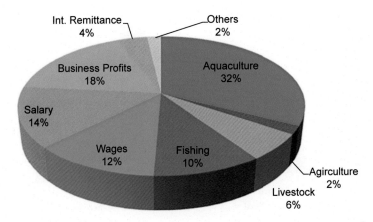

Figure 5.9 Income portfolios of surveyed households.

from the survey data that income distribution across the communities was highly unequal. The bottom one-fifth of the population had a per capita income of BDT8000 per year (US$105) in 2011 and obtained, as a group, only 5% of the total income. In contrast, the top one-fifth of the population attained an average per capita income of BDT54,000 (US$705) in the same year and obtained 54% of the total income.

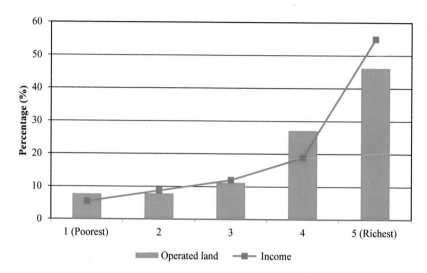

Figure 5.10 Distributions of income and operated land.

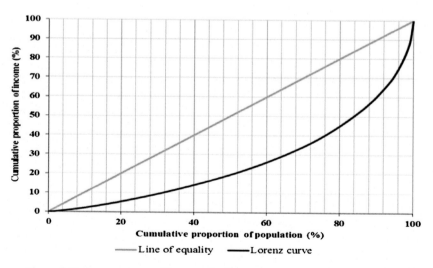

Figure 5.11 Lorenz curves of income distribution among surveyed households.

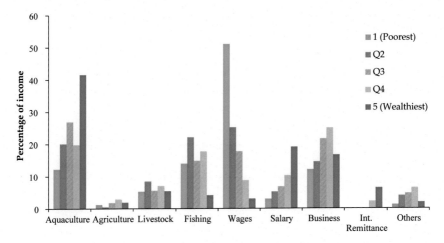

Figure 5.12 Income portfolios across income strata, where Q (quintile) represents 20% of income distribution of all surveyed households.

The importance of different income sources varies across the income groups, as shown in Figure 5.12. The bottom 20% of all households, in terms of income per household, derived a substantial proportion (51%) of their income from nonformal wage employment, whereas the wealthiest 20% of all households supplemented this source with income from shrimp farming and businesses, accounting for 42% and 17% of total income, respectively.

Unlike the poorest and the wealthiest, the middle group did not heavily rely on a particular income source; rather, they pursued livelihood strategies composed of a portfolio of income activities dominated by aquaculture, fishing, wage labor, and small businesses, each of which almost equally contributed to the total households' income. However, across the middle income strata, the proportion of income from wage labor was found to decrease with the increase in the income of household residents, due to the fact that household members of the higher middle class turned to formal wage employment, for which a higher level of education is needed.

5.2.4.2 Loan

In the rural economy, land ownership is an important determinant of access to rural credit from governmental lending institutions. Because of the skewed distribution of land, a small number of medium and large farmers have benefited from subsidized credit from commercial/rural banks, although government resources are insufficient to meet excessive demand.

Table 5.4 Percentage of households by preferred source of credit

Source of credit	Respondents Percentage of n (number of sample) n = 372
NGO	37 (136)
National/private bank	10 (38)
Private money lender	20 (75)
Relatives	23 (85)
Neighbors	3 (11)
Cooperative society	2 (6)
None	6 (21)

Lengthy formalities to sanction loans and speed money are also important hindrances for the rural households. The landless, or functionally landless, that make up nearly half of the rural households have been compelled to turn to NGOs and/or other nonformal lending institutions for small loans. A related issue for the investigation was what type of lending institution people approach to borrow money. This was determined by asking, "If they needed to borrow a large sum of money, who would they turn to for help?"

As shown in Table 5.4, more than one-third of the respondents' preferred option is the NGO's collateral-free small loans. The respondents were asked to point out the reasons for choosing to request loans from NGOs. In reply, 47% of the respondents judged that it was somewhat easy to borrow money from NGOs.

5.2.5 Social Capital

5.2.5.1 Participation in Community Organization

Social capital represents a propensity for collective action for the mutual benefit of the community that derives from the quality of relationships within the community. Active community participants play a pivotal role in shaping an empowered community in which many people are involved in many community activities by forming more of social groups, such as associations, clubs, and voluntary agencies to pursue common goals. The survey data reveals that, apart from NGO organized groups, locally known as *samity*, the participation in social groups or community organizations is low in Chila, where 77% of the total surveyed respondents reported not having involvement whatsoever in any economic, social, or religious groups. Comparing respondents who were engaged in association activities, about 9% of them had an affiliation with national political parties, and 7% reported being members of religious groups such as mosque/temple committees responsible for maintaining mosques or temples—not the general interest of the community. Other local organizations formed by the community on their own volition included youth clubs, Bazzar Committees (representative bodies of the local traders tasked with bringing some sort of order to the local trade), fishing associations, and cooperative associations. Only a little above 3% of the surveyed respondents participated in these types of organizations.

5.2.5.2 Contact with NGOs

The presence of NGOs in the socio-economic sphere is a fact of life in the hazard-prone coastal landscape of Bangladesh. The NGOs are external agents working as catalysts to facilitate the mobilization, participation, and empowerment of disadvantaged communities. They organize landless and poor people, mostly women, into small groups, locally known as *samity*, which act as functional groups with defined set of objectives, such as participatory development and economic empowerment.

Like many other coastal areas, NGO activities in Chila are very much prevalent. During the field survey, a total of 13 NGOs were found to operate in Chila, of which some worked exclusively with coastal issues such as awareness of disaster risk reduction, support for rainwater harvesting; some with broader socio-economic issues, such as micro-credit, gender awareness, and literacy; and the others with both coastal and general issues. NGOs target mainly the poor and socially marginalized households of the society, but it is difficult to estimate how many of them were covered by the NGOs. The major NGO activities in the Chila are:

- Micro credit
- Awareness raising and training on disaster preparedness and management
- Awareness campaign on women empowerment
- Local employment in disaster prevention activities
- Technical training and support for livelihood activities
- Relief and rehabilitation programs

5.3 DISCUSSION AND CONCLUSION

The findings of the rural livelihood analysis, focusing on household assets, provide a detailed overview of the five types of capital assets accessed by rural households across the coastal communities under study. This is done by presenting the results of the analysis of the survey data, capturing the key asset capabilities of the 372 households statistically representative at the union level (lowest administrative level).

5.3.1 Human Capital

The survey data demonstrated that the survey area's human development outcomes are better than the national ones in some respects. For example, overall the literacy rate in Chila is higher than the national rural average rate. Achievement in education can be attributed partly to a higher density of primary and secondary schools in the study area. The average size of household and demographic dependency ratio in Chila is also less than that of the national average. Higher female literacy rates might have contributed to choosing to have fewer children, which, in turn, has resulted in a decreasing age dependency ratio.

However, there are areas with low levels of human capital, such as health and livelihood skills. The survey results reveal that most of the people perceive their health as not being good or very good, for which the scarcity of fresh water in the study area could be the most important reason.

Despite the fact that the provision of technical training is vital for coping with shocks such as cyclones, as well as securing a substantial increase in economic production, a significant number of households in Chila do not have access to livelihood skill training, as found in the household survey. The support of government agencies are constrained by their meager technical services, whereas the NGO sector is well developed in providing a range of technical training, from aquaculture and agricultural production to disaster management, sanitation, and health, as well as tailoring and other small income-generating activities. However, the NGOs' focus is only limited to the poorest and the most vulnerable households in the community, and therefore, a large section of the community members are deprived of access to the skills development training required for improved and climate-resilient livelihoods.

5.3.2 Physical Capital

In the coastal area, a strong dwelling structure is particularly important, as tropical cyclones accompanied by storm surges and wind often lash the coast and sweep away most of the insubstantially built houses and the possessions therein. Ironically, despite the importance of strong housing structures, most of the surveyed households live in small *kacha* houses made of light materials that are locally available. This result reflects the low level of adaptive capacity of the coastal community to cope with extreme climate events.

Pietersen and Beekman (2006) point out that for a healthy life, access to safe water and appropriate sanitation are the most important components. However, like other parts of the southwest region, Chila has a serious problem with drinking water, as the natural sources of drinking water have become contaminated by saltwater intrusion. Faced with difficulties in fresh water availability, households have to compromise with adequate intake of water for their members and place restrictions on using water for cooking, bathing, and other sanitary purposes, especially when they live on purchased water. This is of particular concern because of the health implications for the society. Salt intake through drinking water brings an additional health burden to households, as their salt intake exceeds the recommended limits found in a study conducted in a southwest district (Khan et al., 2011).

With respect to asset ownership, the survey data revealed a relatively poor asset base for the surveyed households in Chila. Among the productive physical assets belonging to households, fishing nets are the most common, followed by fishing boats, whereas households with agricultural equipment are rare, as shrimp aquaculture has mostly replaced agriculture in the study region.

In the coastal belt of Bangladesh, where a severe cyclone on average hits every 3 years, cyclone shelters appear as safe havens, saving the lives of vulnerable people who cannot afford to build strong houses. There are currently 12 cyclone shelters in Chila, but they are not adequate for the population of 200,000, where more than 90% of people live in insubstantially built houses.

5.3.3 Natural Capital

Land is the primary determinant of livelihood in rural Bangladesh. Despite the importance of land in aquaculture and agricultural production as a source of income, four people out of 10 are landless. The distribution of land among landowners is highly skewed, where the top 10% of households own half of the total cultivable land in the study area. Therefore, most of the farmers in the area are marginal or small-farm farmers whose operating farm size ranges between 0.05 and 7.49 acres. The proportion of the total land operated by the large farmers is far more than that prevailing in rural Bangladesh as a whole. The primary reason appearing in the survey is that, confronted with growing salinity in agricultural land, many farmers have partially or fully abandoned agriculture and have leased out or sold land to rich aquaculture farmers and started nonfarm livelihoods.

The survey data suggest that most of households in Chila have relinquished one of the important income-generating activities, cattle raising. Nearly three-quarters of the survey households were unable to keep any ruminants, and the rest maintained a small size of herd because of the lack of animal feed in the area. With increase in soil salinity, crop production has been replaced by brackish water shrimp cultivation, which coincided with the loss of grazing land, thus restraining the traditional home-based livestock farming. This has limited the opportunity for diversification of the income portfolios of households.

5.3.4 Financial Capital

Aquaculture farming practices mostly dominate the rural economy of the southwest coastal area. Despite their predominance, the income from aquaculture/agriculture comprises little more than one-third of annual incomes (40%) of the surveyed households in Chila. Nonfarm income, coming mainly from business, wage and salary employment, and fishing, appear to be important income sources for the communities in this rural coastal area. Across the communities, there are significant income inequalities between the poorest and the wealthiest households in terms of income per capita, as well as income sources. The better-off households tend to be involved in aquaculture and businesses that require large land-holding and investment. On the other hand, the poorest households are likely to be associated with nonfarm wage employment, with little education and few skills.

Turning next to household's access to credit, the NGOs appear to be the most important source for a loan due to the relative ease of borrowing money for most of the households. This is followed by loans from relatives and private money lenders. The survey data suggest that households have limited access to formal money lending institutions.

5.3.5 Social Capital

Active participation is the key to activate social capital, by way of involving citizens in the community's business. The household survey data indicate a high level of nonparticipating community in Chila. This may be correlated with a higher incidence of poverty in the study area, where most of the poor households devoted their time to pursuing livelihood survival strategies. The other reason could be the higher density of NGOs in this vulnerable swathe of land. However, due to their operational approaches, a large number of NGOs working in a number of areas have tended to create social capital by way of creating trust, networks and norms in the community.

Through employment opportunities for locals in their organizations, livelihood skill development and support, and social safety net programs, NGOs have built mutual trust with local communities. For mutual benefit, the NGOs, in many cases, appear to work as a bridge between the local community and the government and donor agencies, to direct resources to vulnerable coastal communities. However, the NGOs' success in creating short-term social capital by providing direct services to the local people has not been achieved without cost; it has come at the expense of community self-reliance and the ability to promote organization within the community (Buckland, 1998).

The survey data presented in this chapter capture the asset capabilities of the households nested in the broader socio-economic situation and vulnerability context of the coastal region. The results show that people in the Chila have a relatively poor asset base, although there are wide differences in the ownership of assets among the community. The asset deficits across most of the coastal communities may be associated with overdependency on, and direct harvesting of, natural resources for livelihoods. This is further compounded by the degradation of freshwater resources as well as by the devastating impact of tropical cyclones. The analysis of the dynamic conditions of the household-level assets and livelihood strategies adopted by the households enable us to understand how households are vulnerable to multiple stressors; on the other hand, the same livelihood assets can result in households' decisions on adaptation to climate factors that eventually affect the sustainable livelihood systems of the households.

REFERENCES

Baert, K., de Norre, B., 2009. Population and social conditions. Perception of health and access to health care in EU-25 in 2007. http://epp.eurostat.ec.europa.eu/cache/ITYOFFPUB/KS-SF-09-024/EN /KS-SF-09-024-EN.PDF (accessed 22.07.13).

Buckland, J., 1998. Social capital and sustainability of NGO intermediated development projects in Bangladesh. Community Dev. J. 33, 236–248.

Khan, A.E., Ireson, A., Kovats, S., Mojumder, S.K., Khusru, A., Rahman, A., Vineis, P., 2011. Drinking water salinity and maternal health in coastal Bangladesh: implications of climate change. Environ. Health Perspect. 119, 1328–1332.

Pietersen, K., Beekman, H., 2006. Freshwater. In: Environmental State-and-Trend: 20-Year Retrospective, United Nations Environment Programme.

World Bank, 2010. Bangladesh: Economics of Adaptation to Climate Change. The World Bank Group.

CHAPTER 6

Local People's Perceptions of Climate Change

Contents

6.1 Introduction	87
6.2 Hydro-Climatic Variability and Extreme Climate Events	88
6.2.1 Temperature and Precipitation Variability	88
6.2.2 Tropical Cyclones Making Landfall Over the Coastal Area	89
6.3 Local Perceptions of Changes in Climate	93
6.4 Comparison between Local Accounts of Climate Change and Meteorological Information	94
6.5 Perceptions of Risks Concerning Climate Change	97
6.6 Discussion and Conclusion	99
References	101

6.1 INTRODUCTION

To be able to make decisions about livelihood adaptation to climate variability and change, households need to interpret and evaluate information about changes in climate parameters. People's subjective judgement of climate change and risk, therefore, is a crucial contributor for allocating efforts by households to address adverse impacts of changing climatic conditions, as has been noted in numerous studies (Thomas et al., 2007; West et al., 2008; Blennow and Persson, 2009). This chapter attempts to answer the following question: What are the perceptions of local communities about historic hydro–climatic variations, change, and extreme events and their corresponding impacts on different livelihood assets, activities, and outcomes?

To answer this question, a useful first step is to characterize the climatic environment to which people in the study area have been exposed in the recent past, which is provided in the first section of this chapter. The next section presents local perceptions of the changes in hydro-climatic variables. In the subsequent section, their narratives of changes in existing climate variables are compared with local hydro–meteorological records for the past several decades. The following section illustrates the particular dimensions of climate variability and extreme events that represent

Experiencing Climate Change in Bangladesh
http://dx.doi.org/10.1016/B978-0-12-803404-0.00006-5

significant disturbances and threats for the societies, shaping local perception of climate risks. This chapter discusses the main findings and closes with conclusions.

6.2 HYDRO-CLIMATIC VARIABILITY AND EXTREME CLIMATE EVENTS

6.2.1 Temperature and Precipitation Variability

Yearly temperatures conditions in Mongla, for a 20-year period from 1989, are presented in Figure 6.1. Temperature records show that winter temperatures in the study area ranged between 12 °C and 32 °C, with maximum temperatures reached in April. The maximum premonsoon summer temperatures fluctuate from a minimum of 32 °C to a maximum of 37 °C.

The rainfall regime of the study area during 1991–2008 is shown in Figure 6.2. During this period, the study area received an annual rainfall ranging from about 1230 mm to 2800 mm. The seasonal distribution of the rainfall was concentrated during the monsoon period (June–September), accounting for 71% of the total precipitation. About 15% of the total rain fell during the summer months (March–May), 12% in the autumn (October–November), and a small proportion (2%) during the winter.

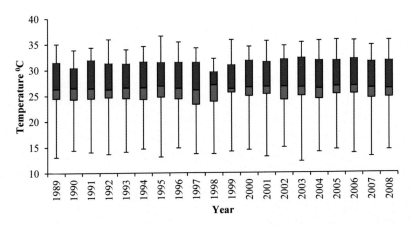

Figure 6.1 Box-and-whisker diagram of yearly temperatures in Mongla, 1989–2008. Whiskers represent maximum and minimum values, and boxes the interquartile range. The solid line within the box is the median. *Data source: Bangladesh Meteorological Department.*

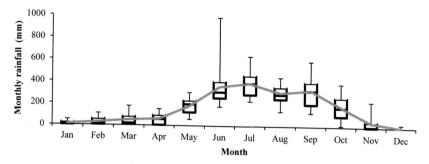

Figure 6.2 Box-and-whisker diagram of monthly total rainfall in Mongla, 1991–2008. Whiskers represent maximum and minimum values and boxes the interquartile range. The solid line within the box is the median and the continuous one, the mean. *Data source: Bangladesh Meteorological Department.*

6.2.2 Tropical Cyclones Making Landfall Over the Coastal Area

The monthly distribution of major cyclonic storms crossing coastal regions of Bangladesh between 1960 and 2013 is shown in Figure 6.3(a). For the period analysed, about 84% of the total tropical cyclones made landfall over Bangladesh in April–May (37%) and October–November (47%), and this temporal distribution of occurrence of tropical cyclones in Bangladesh shows a variable pattern (Figure 6.3(b)).

The trend in the decadal frequencies (an aggregate of 10 years) of the major cyclonic storms over the southwestern coast, in relation to the total of major cyclones over Bangladesh is depicted in Figure 6.4.

The data show that there was a decreasing trend in the frequency of total major cyclonic storms making landfall over the coast of Bangladesh during the three decades from 1960. The rate of decadal decrease was 19% from the 1960s to the 1970s. Between the seventh and eighth decades of the twentieth century, the rate of decrease was 62%; interestingly, in the following decade, the trend reversed toward an increase in cyclones at the same rate (62% per decade). There has been a small decrease in the frequency of total major cyclones in the first decade of the twenty-first century from the previous decade. The interdecadal variation in the frequency of severe cyclonic storms, which have maximum average surface wind speed of 119 km/h or more (WMO, 2013), followed a pattern similar to that of the total cyclonic storms in the past five decades (Figure 6.5). Overall, there is no significant trend in the frequency of the different categories of cyclonic disturbances crossing over the coast of Bangladesh (Tyagi et al., 2009; Wasimi, 2010).

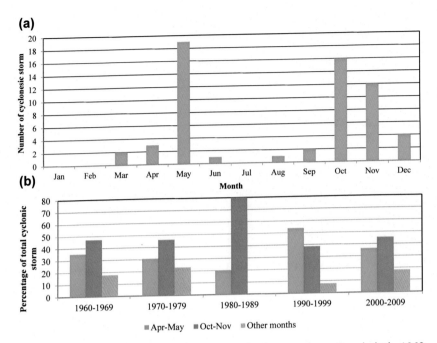

Figure 6.3 (a) Monthly distribution of major cyclonic storms over Bangladesh, 1960–2013. (b) Interdecadal trends in temporal distribution of major cyclone over Bangladesh, 1960–2009. *Data source: WARPO (2013) and EMDAT (2013).*

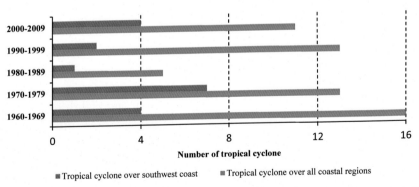

Figure 6.4 Decadal frequency of cyclonic storms over the southwestern coast and all coastal regions of Bangladesh during 1960–2009. *Data source: WARPO (2013) and EMDAT (2013).*

Of the total of 59 major cyclonic storms that struck the coast of Bangladesh between 1960 and 2013, about 31% made landfall over southwestern coastal zone (Figure 6.4). The decadal frequency analysis of the cyclonic storms that struck the burgeoning coastal zone in the past five decades reveals that

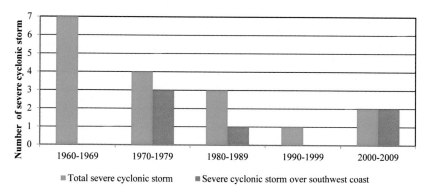

Figure 6.5 Decadal frequency of severe cyclones over the southwestern coast and all coastal regions of Bangladesh during 1960–2009. *Data source: WARPO (2013) and EMDAT (2013).*

there were some differences in the numbers of storms in individual decade, as expected. As indicated by Figure 6.4, the seventh decade of the twentieth century (1970–1979) experienced the highest number (7) of cyclonic storms, whereas in the following decade the corresponding number was 1, which was the lowest occurrence in any decade during the last 50 years. Since then, this trend reversed toward an increased manifestation of cyclonic storms with one in the 1980s and two throughout 1990–1999 to four during the first decade of the twenty-first century. On the other hand, of the total major tropical cyclones that struck southwestern coast, the numbers of severe cyclones throughout 1970–1979, 1980–1989 and 2000–2009 were three, one, and two, respectively, whereas severe cyclone respited in the sixth and ninth decades of the twentieth century.

The maximum wind speed and heights of the storm surges of the major tropical cyclones that crossed over the coast of Bangladesh are shown in Figure 6.6.

A storm surge is the difference between the water level raised by a weather disturbance (storm tide) and the level under a normal astronomical tide. Wind is the main force for generation of storm surges in the way of exerting pressure on the water underneath, resulting a storm wave or storm surge, and the strength of the wind amplitudes the wave. The height of the storm surge varies during various cyclones, ranging from 1 to 6.1 m. Historical data indicate that the frequency of a storm tide (storm surge plus astronomical tide) with a height of about 10 m is approximately once in 20 years, whereas a storm tide of about 7 m occurs every 5 years (MCSP, 1993). The study suggests that the greatest damage in the coastal areas during cyclones has resulted from inundation caused by storm surge. Human causalities have been caused almost exclusively by storm surges (Khalil, 1992; Dasgupta et al., 2010).

Figure 6.7 shows the total number of casualties per decade associated with major tropical cyclones in low-lying coastal areas in Bangladesh during the last 50 years to 2009. The most striking feature of the impact of these cyclonic storms has been the massive human casualty. During the past 50 years, major cyclones claimed about 530,000 human lives.

Indeed the number of deaths is very much dependent on single extreme events, as were the cases for the deadliest cyclones of 1970 and 1991. In November 1970, Bangladesh coast experienced the deadliest tropical cyclone in history, in terms of human losses. This cyclonic storm was classified as a category 4 event and made landfall over the Bhola-Meghna estuary, with the wind speed of about 222 km/h, claiming approximately 300,000 lives. In 1991, the Noakhali-Chittagong coastal area was hard hit by a very strong

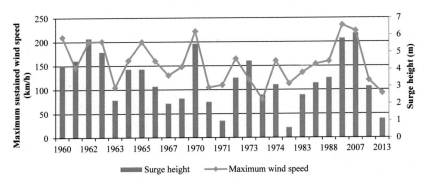

Figure 6.6 Maximum wind speeds and corresponding surge heights of the major cyclonic storms during 1960–2013. *Data source: WARPO (2013) and EMDAT (2013).*

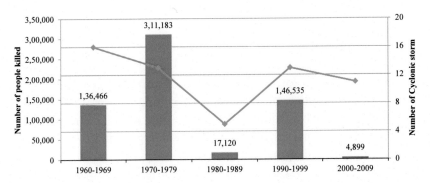

Figure 6.7 Decadal frequency of major cyclonic storms and associated human casualties during 1960–2009. *Data source: WARPO (2013) and EMDAT (2013).*

tropical cyclone (category 4 event) with wind speeds of 235 km/h and a storm surge of 4–8 m, killing between 130,000 (BBS, 1991) and 200,000 (Rashid, 1991) people. During a more recent period, Bangladesh has managed to reduce the number of human fatalities related to cyclonic storms. For example, the tropical cyclone Sidr made landfall over the Sundarban-Barguna coast in 2007 (a category 4 event). The sustained wind speed was approximately 223 km/h, and the average surge heights ranged between 6 and 8 m. The fatality figures caused by the cyclone Sidr were 4275, approximately a 99% reduction compared with the deadliest 1970 cyclone.

6.3 LOCAL PERCEPTIONS OF CHANGES IN CLIMATE

Respondents in the study area described changes in climate trends around them by describing their personal experiences of changes in seasons, weather patterns, and sea level rises. They related the changes in long-term patterns in the climate to increased extreme weather events, variability in the weather and climate parameters, and impacts on their local physical environments. Table 6.1 shows the local views of changes in climate characteristics made by the participants in the focus group discussions.

Table 6.1 Local perceptions of climate change

		Local experience of changes
Climatic variables	Temperature	• Summer hotter than before • Increase in warmer days • Increase in frequency of abnormally hot days/heat weaves
	Rainfall	• Decrease in total rainfall in rainy season • Total rainfall decreased in winter season • Number of annual rainy days decreased
Climatic phenomena	Monsoon	• Shorter rainy season • Late onset of rainy season
	Tropical cyclones	• Increase in tropical cyclonic activities • Increase in height of storm surge and wind velocity • Increase in area flooded by cyclonic storm surges
Impacts on physical environment	Sea level rises	• More area being inundated during high tides • More area being intruded on by salinity • Increase in the level of salinity during summer time

There were many ways in which respondents described their observed changes in the patterns of seasonality, of which the most common changes experienced included warmer summers characterized by more intense heat, drier winters with decreasing rainfall, and shorter monsoon seasons with decreasing trends in total rain in this season.

Changes to cyclonic storms were described by the respondents, which were related to an increased occurrence of tropical cyclones, with strong winds and high storm surges resulting in extensive inundation of the coastal area.

Some respondents observed that the high tidal waters reached areas along the coasts, estuaries, and coastal rivers that were previously above the normal high tide levels. This change was quite evident to respondents who live outside the embankment and behind the shore as they experienced the high tides that now reached their homes, causing land erosion of homesteads. Respondents were less likely to perceive this change as a manifestation of sea level rise due to global warming. They predominantly attributed the change in tidal inundation to an increased sedimentation and siltation of the rivers, and restricted river flows resulting from embankments built for shrimp farming. Very few participants mentioned sea level rise as the cause of increasing high water.

Increasing salinity intrusion from the Bay of Bengal is a well-known phenomenon in southwestern coastal districts. As expected, all the respondents reported an increase in salinity encroachment and the level of salinity during the dry season. While forming attribution for the increasing saltwater intrusion, most of the respondents described it as a result of cyclonic storm–induced coastal flooding, prolonged brackish water shrimp cultivation, freshwater withdrawal from upstream in the rivers, and low seasonal rainfall.

6.4 COMPARISON BETWEEN LOCAL ACCOUNTS OF CLIMATE CHANGE AND METEOROLOGICAL INFORMATION

The respondents' descriptions regarding their experiences of changes in climatic characteristics were evaluated to ascertain the degree of correlation between people's perceptions of changing climate and the scientific evidences of observed hydro-meteorological trends of the study region.

The meteorological records for the period between 1980 and 2007 for all 34 stations in Bangladesh showed that there were increasing linear trends in seasonal and annual mean temperatures since 1980, as reported by the Climate Change Cell of the Department of Environment in Bangladesh (Climate Change Cell, 2009). The Climate Cell reports that rising trends in annual and seasonal mean maximum temperatures have been observed since the 1970s or 1980s, depending on the region in question. Since 1980, the rising trends for the annual, winter (November–February), summer (March–May), and monsoon (June–October) seasons were 2.14, 1.33, 2.15, and 2.44 °C per century, respectively. Similar trends are observed in mean maximum temperatures for all months except January (peak winter month), which shows cooler trends. From May to September, the warmer trends have been statistically significant since 1980. As mean temperature has increased strongly since 1980, it is very likely that the respondents would have experienced hot temperature extremes in summer and monsoon seasons with higher frequency. Thus, there is a strong correlation between the respondents' perceptions of increasing warming weather and temperature data.

The ways in which respondents reported that they had experienced changes in rainfall were: a decrease in monsoon and winter rainfall and a decrease in number of rainy days annually. To verify their claims, the respondents' perceptions of changes in precipitation were compared with rainfall data from a local meteorological station. Neither the winter nor the seasonal rainfall time series for 18 years (1991–2008) show any statistically significant changes over this time period. However, the Climate Change Cell study shows that, at a regional scale over the period of 1981–2001, monsoon rainfalls appeared to have decreased in the central and southern part of the country but increased in the northern part (Bogra, Mymensingh, and Sylhet areas), compared to the reference period of 1960–1980. Winter rainfall exhibited an increasing trend for the same period all over the country, except in the northeastern part (Sylhet area). However, some other studies do not support the claim of decreasing rainfall in the southwestern part of the country during monsoon season (Shahid, 2009; Mondal et al., 2013).

At the next level, the analysis is aimed at investigating rainfall anomalies from the Mongla station's 1991–2008 mean and 5-year running averages. During the 18-year period, as represented in Figure 6.8, no convincing pattern in monsoon rainfall anomalies could be noted in Mongla, in comparison to the 1991–2008 mean. However, the 5-year moving average of monsoon rainfall shows that, except for 2002 and 2005, the trend of below-average rainfall has persisted since 2000, supporting the respondents' perceptions of a decline in monsoon rainfall.

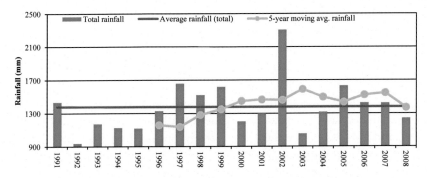

Figure 6.8 Annual departure of monsoon rainfall, 1991–2008 mean.

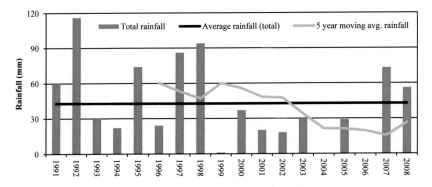

Figure 6.9 Annual departure from winter rainfall, 1991–2008 mean.

Some winter rainfall recovery in 2008 and 2009 is apparent in Figure 6.9, but the winters in 2004 and 2006 were rainless, and that might capture the respondents' attention and lead them to believe in decreasing winter rainfall. Whether the number of rainy days is decreasing annually was not investigated because of the unavailability of daily rainfall data from the local meteorological stations.

The respondents perceived that the rainy season now begins late, but does not flow into the next seasons and therefore accounts for the shortened length of the monsoon season. The rainy monsoon season in Bangladesh starts in June and lasts until October. However, the period from mid-October through to mid-November is also considered as autumn. The personal experience of late occurrence of the rainy season may be due to the decline in the number of rainy days in the month of June, the first month of rainy monsoon. Unfortunately, monthly rainfall records for Mongla station, instead of daily or at least 10-day rainfall data, did not allow us to test whether the numbers of rainfall events have decreased in June. However, it was revealed from the monthly trend analysis of the rainfall data from the adjacent meteorological station that the rainfall in June significantly decreased (Mondal et al., 2013).

The residents' accounts of tropical cyclones becoming more frequent and more intense, with increased storm surge heights and more spatial inundation, may be because of the occurrence of two severe cyclones in the study area in the recent past. From 1970 to 2011, there were 18 major storm surges striking the southwestern coastal areas, of which 6 were severe cyclonic storms. Within the 5 years before this survey, respondents encountered two of the most severe storms that they had experienced in almost two decades. The cyclonic storm that hit the study region in 2007 was a severe storm surges, with wind speeds of 223 km/h and a storm surge of 6.1 m. A year had passed since the residents in the country's southwestern coast faced another severe cyclonic storm in 2009, with maximum wind speeds of 120 km/h and tidal surges of up to 3 m.

The respondents' observations concerning the salinity intrusion into nonsaline areas and the rising level of salinity in soil and water is consistent with the soil survey and water quality monitoring records, in which observed changes in both soil salinity class and level of salinity toward an increasing trend have been reported (SRDI, 2010; Rahman et al., 2011). The change observed by the respondents with regard to expansion of the area inundated due to rising high tide water levels has been confirmed by some recent studies (World Bank, 2010; Mondal et al., 2013; Brammer, 2014).

6.5 PERCEPTIONS OF RISKS CONCERNING CLIMATE CHANGE

While attempting to understand the respondents' perceptions of the major risks associated with climate variability and change, the respondents in the group discussions identified some of the disturbances that can be explicitly attributed to climatic factors, whereas others are related to the impacts of climate change entwined with nonclimatic factors, such as salinity intrusion. Livelihoods in the study area are predominantly based on aquaculture, dominated by shrimp culture. Because of this particular farming activity, the respondents were found to attach significance to those risks that they experienced as the significant contributors to the crises in shrimp farming. Table 6.2 shows the characteristics of climate change perceived by the respondents as major risks because of their potential to adversely affect local livelihoods.

Table 6.2 Local perceptions of risk of climate change

Perceived changes in climate	Key impacts
Increase in summer temperatures	• Public/animal health problem • Disease outbreak/mortality in shrimp farms • Increase in salinity due to high evaporation
Later onset of rainy season Decline in seasonal rainfall Increase in cyclonic activities	• Stocking of freshwater fish delayed • Increase in salinity • Injuries and loss of lives • Loss of crop, fishstock, livestock, and poultry • Damage to public and private infrastructure • Contamination of drinking water sources • Impacts on livelihood activities
High tide rising	• Salt-line moves to inland • Land erosion

Most of the respondents identified increased temperatures in the summer months, with an increase in the frequency of unusually warm days, as a risk because it causes human and animal health problems, damage to aquaculture production, and increased salinity in the various natural water sources from which drinking water is obtained by the local residents. High water temperature causes a reduction in dissolved oxygen levels in the shrimp pond water, affecting shrimp growth and survival. Prolonged exposure to low oxygen levels often contributes to disease outbreaks and the mass mortality of shrimp stock (FAO, 2013). The evaporation rate from shallow shrimp ponds increases due to intense heat in the summer, causing high levels of salt concentration in the ponds. The respondents noted that the exposure to the stress of heat and high salinity resulted in the retarded growth and the mortality of shrimp.

The late onset of the monsoon rainy season, marked by erratic rainfall patterns at the beginning of the season, was viewed as a risk, as it affects the farm production cycle. After harvesting the shrimp and euryhaline fish species, which are cultivated during the high-salinity period, shrimp farmers expect monsoon rain to flush the salt from the shrimp ponds at the end of this production cycle. When monsoon rain causes lower salinity levels in the pond environments, freshwater fish and prawns are cultured together with brackish-water shrimp during July and December. People say that the onset of rainy season now occurs later in the season, causing delays in stocking freshwater fish.

As discussed above, tropical cyclones disrupt every aspect of society on the coast, and shrimp aquaculture is no exception. Therefore, all of the participants were concerned about the vulnerability of shrimp farms to cyclones accompanied by storm surges due to ineffective coastal embankments or a lack of flood control structures. When cyclones hit the area with increased heights of storm surges, the surge water breaches the earthen embankments encircling the ponds and inundates the shrimp ponds, allowing shrimp to escape into flood plains and water courses.

Saltwater intrusion is a major concern in the shrimp farming regions because it has degraded the coastal ecosystems by causing soil degradation, deforestation, destruction of homestead vegetation, loss of coastal vegetation, and a decline in agro-biodiversity (Rahman et al., 2011; Paul and Vogl, 2011). With the increasing salinity in surface and ground water, freshwater becomes scarcer, and that, in turn, affects economic and household activities.

In the Mongla area, people living in the proximity of estuaries or rivers and outside the coastal embankment are exposed to rising high tide causing soil erosion and salinity intrusion. However, the high tide rise is not perceived as a major risk, especially by the people living in the areas protected by coastal embankments.

6.6 DISCUSSION AND CONCLUSION

In the southwestern coastal region of Bangladesh, the long-term rise in temperature reported in a number of scientific reports closely conforms to the beliefs held by most respondents. Similar correlations have been found between local observations and the meteorological records about changes in high tidal water levels and salinity. Respondents also reported experiencing climate change in terms of changes in winter rainfall, cyclonic activity, storm height, and wind speed, but these are not significantly substantiated by the scientific evidence. However, when compared with the aforementioned trends, based on the short-term records, local perceptions show a strong correlation with the meteorological records. This reflects the difficulties in detecting trends by personal experience within the random natural variability in conditions that can be expected for both stable and changing climates (Weber, 2010).

In line with previous findings, this study has found that recent memories of climatic events could affect local people's beliefs in the direction of change. For example, the repondents' faulty judgments about changes in cyclonic activities could be linked to two recent cyclones within a short interval (2007 and 2009), which would normally occur farther apart. In a study among a Papuan population in Indonesia, Boissiere et al. (2013) reported that people's beliefs in increased frequency of flood was associated with two recent floods in the study area, showing a correlation between the local interpretation of climatic trends and recent climatic events.

In the study area, the most frequently mentioned climatic characteristics causing risk to life and livelihoods are associated with extreme climate events, notably intense heat and increased cyclonic activities; changes in patterns in rainfall characterized by the late onset of rainy season and a decline in seasonal rainfall; and the impacts of climate change on physical environments, such as rising tidal water levels and salinity intrusion. It appears that the adverse impacts of climatic events on aquaculture play an important role in shaping people's risk perceptions of climate change, although they affect other spheres of life too. The underlying reason for this association is understandable; aquaculture dominates the rural economy in the southwestern coastal region, and this farming practice is very sensitive to changes in hydro–climatic conditions. Shrimp farming thriving on the southwestern coast in the favorable hydro-climatic conditions of this area is resilient to the normal interannual variability of environment. When the required hydro-climatic conditions for aquaculture exceed the normal range, farmers experience adverse impacts on fish production. Whether these altered conditions reflect short-term variability or long-term change, they appear to local farmers as indicative of climate change. Weber (2010) argues that repeated personal experiences of hazardous weather events with adverse consequences lead to physical evidence of climate change in an individual's perception, which in turn results in concern about climate change, i.e., perceptions of its riskiness.

People on the southwestern coast have experienced the increasing impacts of hydro–climatic variability and change. Over the years, repeated experience with negative consequences resulting from climatic events on livelihood activities, principally on shrimp aquaculture in the context of the study region, has shaped people's perceptions of climate risk. The results of the study show that local accounts of climate change mostly diverge from scientific evidence when we consider long-term climate trends; but on short-term variability, the correlation between the scientific and local accounts is high.

People in the case study area perceive that the hydro–climatic environment in the southwestern coastal region has already changed, which, on their accounts, is manifested by rising temperatures, declining rainfall, increasing cyclonic storms, growing salinity intrusion, and increasing high tide levels. When local accounts of directional changes in the climatic parameters are compared with meteorological records, the correlation between the views of the meteorological experts and the local communities is not unequivocal. The research reveals that the communities' perceptions concerning the long-running trends in climatic variables, such as rainfall or climatic phenomena (e.g., tropical cyclone) diverge from scientific understanding, whereas the local accounts of climate change impacts on the physical environment closely matches the scientific evidence.

REFERENCES

Blennow, K., Persson, J., 2009. Climate Change: motivation for taking measures to adapt. Global Environ. Change 19, 100–104.

BBS, 1991. Statistical Yearbook of Bangladesh 1991. Bangladesh Bureau of Statistics, Bangladesh.

Brammer, H., 2014. Bangladesh's dynamic coastal regions and sea-level rise risks. Clim. Risk Manage. 1, 51–62.

Boissière, M., Locatelli, B., Shiil, D., Padmanaba, M., Sadjudin, E., 2013. Local perceptions of climate variability and change in tropical forests of Papua, Indonesia. Ecol. Soc. 18 (4), 13.

Climate Change Cell, 2009. Characterizing Long-term Changes of Bangladesh Climate in Context of Agriculture and Irrigation. Climate Change Cell, Department of Environment, Bangladesh.

Dasgupta, S., Huq, M., Khan, Z.H., Ahmed, M.M.Z., Mukherjee, N., Khan, M.F., Pandey, K., 2010.Vulnerability of Bangladesh to in a Changing Climate.World Bank Policy Research Working Paper 5280.

FAO, 2013. Shrimp Culture: Pond Design, Operation and Management. FAO training series,Version-2, p. 76 http://www.fao.org/docrep/field/003/AC210E/AC210E09.htm (accessed 20.03.13).

Khalil, G.M., 1992. Cyclones and storm surges in Bangladesh: some mitigative measures. Nat. Hazards 6, 11–24.

MCSP, 1993. Summary Report: Multipurpose Cyclone Shelter Project. Bangladesh University of Engineering and Technology. Bangladesh Institute of Development Studies, Dhaka.

Mondal, M.S., Jalal, M.R., Khan, M.S.A., Kumar, U., Rahman, R., Huq, H., 2013. Hydro-meteorological trends in southwest coastal Bangladesh: perspective of climate change and human interventions. Am. J. Clim. Change 2, 62–70.

Paul, B.G.,Vogl, C.R., 2011. Impacts of shrimp farming in Bangladesh: challenges and alternatives. Ocean Coastal Manage. 54, 201–211.

Rashid, H., 1991. Geography of Bangladesh. University Press, Dhaka.

Rahman, M.H., Lund, T., Bryceson, I., 2011. Salinity effects on food habits in three coastal, rural villages in Bangladesh. Renewable Agric. Food Syst. 26 (3), 230–242.

Shahid, S., 2009. Rainfall variability and the trends of wet and dry periods in Bangladesh. Int. J. Climatol. 30, 2299–2313.

SRDI, 2010. Saline Soils of Bangladesh. Soil Resource Development Institute, Bangladesh.

Thomas, D.S.G.,Twyman, C., Osbahr, H., Hewitson, B., 2007. Adaptation to climate change and variability: farmer responses to intra-seasonal precipitation trends in South Africa. Clim. Change 83, 301–322.

Tyagi, A., Mohapatra, M., Bandyopadhyaya, B.K., Kumar, N., 2009. Inter-annual variation of frequency of cyclonic disturbances landfalling over WMO/ESCAP Panel Member Countries. In: Al-Hatrushi, S.,Yassine, C. (Eds.), 1st WMO International Conference on Tropical Cyclones and Climate Change. Muscat, Oman, pp. 1–7.

West, C.T., Roncoli, C., Ouattara, F., 2008. Local perceptions and regional climate trends on the Central Plateau of Burkina Faso. Land Degrad. Dev. 19, 289–304.

WMO, 2013. Terminologies Used in the Region of the Bay of Bengal and the Arabian Sea: Severe Weather Information Centre. World Meteorological Organization. http://severe.worldweather.wmo.int/tc/in/acronyms.html #SCS (accessed 09.09.13).

Wasimi, S.A., 2010. Statistical forecasting of tropical cyclones for Bangladesh. In: Charabi, Y. (Ed.), Indian Ocean Tropical Cyclones and Climate Change. Springer, USA, pp. 131–141.

World Bank, 2010. Bangladesh: Economics of Adaptation to Climate Change. The World Bank Group.

Weber, E.U., 2010. What shapes perceptions of climate change? WIREs Clim. Change 1, 332–342.

CHAPTER 7

Climate Disturbances and Change: Strategies for Adaptation

Contents

7.1	Introduction	103
7.2	Livelihood Diversification for Adaptation and Increasing Security	104
7.3	Changing Livelihood Strategies for Adaptation to Climatic Hazards and Other Stressors	106
	7.3.1 Shift to Aquaculture-Based Livelihood Strategies	107
	7.3.2 Incorporation of Traditional Practices into Commercial Aquaculture	108
	7.3.3 Adoption of New Species as a Risk-Spreading Strategy	109
7.4	Coping Strategies in Shrimp Aquaculture	110
7.5	Adaptation to Salinity Intrusion in Rice Production	111
7.6	Use of Climate Information	113
7.7	Adaptation to Salinity Encroachment in Drinking Water Resources	114
7.8	Improvement of Shelters: Households' Response to Tidal Flood	115
7.9	Migration	116
7.10	Discussion and Conclusion	117
	References	121

7.1 INTRODUCTION

This chapter illustrates the types of mechanisms by which livelihood adaptations are occurring at the household level in the context of coastal regions. Having analyzed in previous chapters the adaptive capacity in terms of households' access to five main types of assets and the local understanding of the livelihood risks associated with climate change, this chapter examines how households adapt their livelihoods to respond to climate risk. This chapter explores how the adaptive capacity of the households, triggered by climate risk perceptions, enables the process of adaptation to climate change. Here livelihood adaption to climate change is regarded as incremental adjustment in the routine livelihood strategies or transformation of livelihood systems to withstand and recover from short–term weather–related shocks and also changes in long–term climatic conditions.

This chapter explores adaptation and livelihood diversification, first by examining in detail the potential for diversification of household income

Experiencing Climate Change in Bangladesh
http://dx.doi.org/10.1016/B978-0-12-803404-0.00007-7

sources as an adaptation option to reduce vulnerability to climate related income-loss. Second, the chapter considers changing livelihood activities as a response strategy used by households in response to the changing hydro-climatic conditions within the prevailing socio-economic and political contexts. Third, the chapter discusses the specific adaptation measures in response to climate variability and the changes being implemented by agriculture and aquaculture farmers. The final section of this chapter discusses the main findings of the study and draws conclusions.

7.2 LIVELIHOOD DIVERSIFICATION FOR ADAPTATION AND INCREASING SECURITY

Livelihood diversification of rural households has long been an important adaptation option in many countries. Ellis (2000) argues that rural households in developing countries diversify their livelihoods due to insufficient income from a single livelihood activity, for example, if farming on their own land does not provide sufficient means for the survival of many rural households. Livelihood diversification has the potential to spread risk over a portfolio of activities and to minimize the vulnerability of households in the event of a failure in the major source of income due to exposure to risk factors (e.g., tropical cyclones), and this is compensated for by the complementary income streams of the households (Ellis, 2000; Paavola, 2008). Figure 7.1 illustrates the diversification of livelihood activities across income groups in Chila, as revealed in the household survey.

Households pursue a diverse set of activities and income sources as a livelihood strategy for the well-being of their families. The survey data show

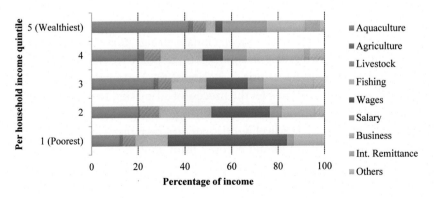

Figure 7.1 Income portfolios across income strata.

that the bottom 20% households, in terms of income per household resident, derived a substantial proportion (51%) of their income from nonformal wage employment, whereas the wealthiest 20% of households supplement this source with income from shrimp farming, salaried jobs, and businesses, accounting for 42%, 19%, and 17% of their total income, respectively.

Unlike the poorest and the wealthiest, the middle group does not heavily rely on a particular income source; rather, they pursue a livelihood strategy composed of a portfolio of income activities dominated by aquaculture, fishing, wage labor, and small businesses, each of which is found to be almost equally contributing to the total households' income. However, across the middle income strata, the proportion of income from wage labor was found to decrease with an increase of the income of the household residents. This was found to be due to the fact that household members of the higher middle class turned to formal wage employment, for which a higher level of education is needed.

Observed livelihoods could be better understood when the livelihood strategies of the households are categorized according to the size of the landholding operated by the household. Figure 7.2 shows the portfolio of activities among the surveyed farming and nonfarming households. The survey data explicitly reveal that land-holding plays a major role in a household's decisions about livelihood strategies. It is a given that a large farm holding is associated with a higher income, as illustrated in Section 5.2.4 (Figure 5.10), and therefore, for a large-farm farmer, investment in shrimp aquaculture was seen to be a good way for an individual household to increase its income and save money to invest in livestock farming, nonfarm businesses, and giving children a better education to improve their prospects of getting salaried jobs.

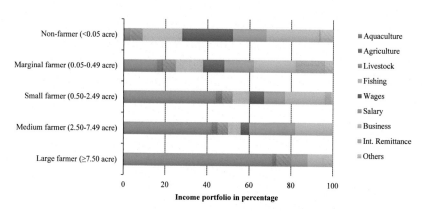

Figure 7.2 Income portfolios across farming households.

At the time of the survey, aquaculture made up a little above 40% of the total income of medium and small farmers, and for marginal and nonfarmers, aquaculture's contribution was proportionately smaller in accordance with their land-holdings. Nonfarmers are functionally landless, but they were found to cultivate shrimp in their homestead ponds. The survey data show that with decreasing income from on-farm activities, households were increasingly diversifying their nonfarm income sources as their livelihood strategies to increase their total income, as well as to reduce potential threats to their livelihood security due to the various risk factors.

It also appears from the survey data that the households pursue similar strategies for their livelihoods but differ in the share in major sources of income. Better-off households predominantly depend on shrimp aquaculture as their primary income stream, although shrimp farming is more likely to involve high risks from diseases, international markets, supply of shrimp fry, tropical cyclones, and climatic hazards. However, these households tend to reduce total income risk by having in place a less risky cash flow from businesses and salaried jobs. On the other hand, middle-income households appear to be less vulnerable to the impact of climate on their livelihood security, as climate-sensitive on-farm income accounts for 21–29% of their total income. They have diversified a portfolio of nonfarm activities that results in risk spreading with a comparatively small loss in total household income while facing adverse climatic impacts. Poor households pursue survival livelihood strategies due to the fact that their income-earning opportunities are limited, as they lack access to livelihood assets (see Chapter 4). Households in this lowest income stratum were overly reliant on wage labor as a means of survival in this rural setting, making up approximately 51% of their total income. Their off-farm wage labor (wage labor on other farms) was found to have a direct correlation with the risks associated with agricultural and aquaculture production in the area. When there are weather-related hazards, such as tropical cyclones, heavy rainfall, and intense heat in the area, almost all of the income sources of the poor households are the hardest hit.

7.3 CHANGING LIVELIHOOD STRATEGIES FOR ADAPTATION TO CLIMATIC HAZARDS AND OTHER STRESSORS

Local perceptions of changes in climatic characteristics and their impacts causing a risk to livelihoods were discussed in Chapter 6. Confronted by these livelihood-affecting risks within the wider rural socio-ecological

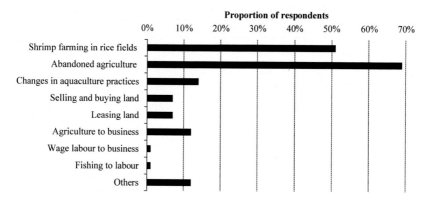

Figure 7.3 Changing patterns of livelihood activities over the last 10 years in Chila.

changes, households have adapted to these changes by deliberately adjusting their livelihood activities in order to exploit beneficial opportunities or to reduce vulnerabilities arising out of these changes. Figure 7.3 illustrates how the livelihood strategies in Chila have changed over the last 10 years in response to the economic circumstances and environmental conditions.

The survey information shows that the most significant change in livelihoods was brought about by shifting the economic production system from agriculture to brackish water shrimp aquaculture. About 70% of the total surveyed households with agriculture in their farming portfolio decided to abandon agriculture, and of which approximately 70% chose to invest in shrimp aquaculture and approximately 20% in businesses during the last 10 years.

7.3.1 Shift to Aquaculture-Based Livelihood Strategies

Traditionally, households in Chila mostly relied on rice farming, fishing, and gathering forest materials from the Sundarban Forest to secure their livelihoods. On the basis of agriculture-based economic production systems, households diversified their income portfolio by integrating subsystems such as family poultry, goat grazing, and livestock herding, etc. Shrimp aquaculture began to expand in Chila in the mid-1980s. By taking advantage of the numerous rivers and tributaries and low-lying tidal flood plains, brackish water shrimp aquaculture has undergone a rapid horizontal expansion due to an increased demand in the international market, high prices, and assistance through international donor-funded projects (World Bank and Asian Development Project) and government incentives, such as amendment of the land lease laws, income tax rebates, and subsidized credit (USAID, 2006; Swapan and Gavin, 2011; Afroz and Alam, 2013).

With the rapid expansion of shrimp aquaculture, farmers have started to experience growing levels of salinity in their cultivable lands in which brackish water shrimp culture is one of the factors causing the salinity and affecting crop production to a great extent (Deb, 1998; Ali, 2006). The contributing factors for salinity intrusion and its seasonal variation have been discussed in Chapter 6. Depending on the degree of salinity in the cultivable land and the salt-tolerance capacity of the crop, salinity reduces crop yields or causes total yield loss. Agricultural production being constrained by increased salinity ingression, especially in the dry season, causes rural people to respond to the increase in salinity by expanding gher farming into agricultural land previously dedicated to rice production.

The shift to more salt-resistant farm practices as a primary economic production system of the households entails changes in livelihood portfolios, comprising its subsystems. In the study region, the decline in rice production coincides with a collapse in households' traditional sources of income ranging from poultry and livestock to homestead vegetation; however, it increases income opportunities from the subsystems of shrimp farming that include fry collection, shrimp nurseries, trading, processing, and exporting.

7.3.2 Incorporation of Traditional Practices into Commercial Aquaculture

In Chila, farmers have introduced changes in the farm production system to reduce exposure to hydro-climatic risks and to increase the flexibility of farm production. In the newly developed aquaculture-based economic production systems, mud crab culture appears to be one of the more specific strategies adopted by farmers to reduce their vulnerability to salinity—the main hydro-climatic concern in the region. Mud crabs can survive in high salinity (5–40 ppt) and high turbidity, and have good growth in high temperatures (Shelley and Lovatelli, 2011). Traditionally, people in the southwest coastal districts harvest mud crabs from swamps in the Sundarban Forest, estuaries, rivers, and canals for livelihood support; and these wild mud crabs account for 90% of the total exported crabs in Bangladesh (Zafar and Hasan, 2006). Now this traditional practice of direct collection of natural resources is being transformed into a commercial aquaculture production system. In many coastal districts in Bangladesh, both monoculture and polyculture farming are practiced in mud crab cultivation. In polyculture systems, seed crabs are stocked with shrimp and other fish species in shrimp gher, which are then reared for 6–8 months to become a marketable size. The monoculture system, on the other hand, is a short-term rearing of immature, or recently moulted, crabs until maturity that generally lasts for 1–2 months. In this system, only crabs are cultivated for fattening, in ponds fenced with bamboo screens and nylon nets to prevent them from escaping, as well as in pens and cages. To supply seed crabs for farms, crab hatcheries have started to emerge in the coastal region as a new business opportunity, and a few of them were found operating in Chila. Despite the potential of being an adaptation option that is less sensitive to high levels of salinity and high temperatures, crab farming is still in its early phase, with only 2% of the surveyed farmers having attempted crab farming. One of the underlying reasons may be religious, as most of the Muslims in Bangladesh do not eat crab, and therefore it was traditionally limited to the minority Hindu community.

7.3.3 Adoption of New Species as a Risk-Spreading Strategy

Experimentation in aquaculture practices has been a long tradition in the southwestern coastal region in Bangladesh. To optimize livelihood outcomes, well-off farmers began commercial shrimp farming experimentally in the early 1970s, but it has undergone a more rapid expansion than any other agro-export commodity in Bangladesh. Later, during the late 1970s and the mid-1980s, local farmers experimented with, and developed, freshwater prawn-farming systems in rice fields in low-lying agricultural land (Belton et al., 2011). As with most local innovations, farmers are adopting new fish strains into their aquaculture farming systems in response to perceived risks attached to shrimp cultivation. Research participants in Chila noted that the income from shrimp cultivation could be declining significantly due to disease outbreaks or extreme weather, leading to severe, and sometimes permanent, detrimental impacts on the livelihoods systems of the shrimp farmers. Whether to reduce production risk or increase productivity levels, the most common adaptations incorporated into aquaculture practices in the last 10 years in Chila were mud crabs and mono-sex tilapia (all male species), representing 14% of the total surveyed households. The salinity tolerance capacity of the mono-sex tilapia species opens up the opportunity for farming in brackish water aquaculture farming systems in the coastal production system. In Chila, farmers reported farming mono-sex tilapia within polycultures, with white fish (freshwater fish) and shrimp in shrimp ponds. Although shrimp and prawns are produced primarily for the foreign market, tilapia and other white fish are marketed for local consumption.

7.4 COPING STRATEGIES IN SHRIMP AQUACULTURE

Table 7.1 shows the specific strategies that shrimp farmers used to address interannual variability and the other generic adjustments made in their production strategies in the study area.

In Chila, shrimp farmers have adapted to flooding by investing in farm infrastructure. Of the respondents surveyed, 47% reported that they had increased the embankment heights around the shrimp ponds prevent fish from escaping during inundation caused by intense rain in short durations or tidal floods. To make this intervention more effective, 3% of respondents placed nets around the shrimp gher so that the shrimp farm could sustain destructive flooding with increased water levels.

As discussed in Chapter 5, farmers perceived that there was an increase in the number of summer days with intense heat that accounted for the decline in shrimp production. Understanding the relationship between the exposure to the stress of heat and to the retarded growth and the mortality of the shrimp, 20% of respondents dug deep ponds inside the shrimp fields so that the shrimp could take refuge to escape from heat stress during the summer season.

Aside from the above-mentioned strategies being used as specific responses to climatic events, some farmers have adopted other, new techniques to maintain healthy aquaculture environments in their shrimp ponds, which can be adversely affected by high temperatures and intense rainfall. Of these new practices, 10% of the surveyed farmers reported applying lime to manage the soil and water quality of shrimp ponds, and 3% used medicine for treatment

Table 7.1 Adaptation measures implemented by shrimp farmer in Chila

Changes in shrimp farming practices	Respondents' percentage of n (number of sample) n = 30
Increased embankment height	47 (14)
Digging ponds inside the fish farm	20 (6)
Liming	10 (3)
Using medicine	7 (2)
Placing nets around the shrimp fields	3 (1)
No response measures implemented	47 (14)

of shrimp diseases. Data suggest that there were significant numbers of people (47%) who did not implement any measures against climatic or non-climatic hazards affecting shrimp cultivation.

In attempting to understand the reason, respondents were asked whether they were worried about the climatic impacts and the reasons for their inaction. As expected, most participants were concerned about climate change and extreme events, but they could not afford to take adaptation measures.

As shown in the survey data and the perspectives that the farmers articulated through group discussions, the vulnerability of shrimp farming to climate change is tightly coupled with environmental problems, under-development, food insecurity, and socio-economic inequalities. Shrimp farmers, especially the lower segments of the shrimp value chain, have been trapped in chronic poverty that limits their ability to respond to climatic risks.

7.5 ADAPTATION TO SALINITY INTRUSION IN RICE PRODUCTION

Rice production in the study region has been affected severely by increasing soil salinity. With the expansion of shrimp cultivation in fields previously devoted to rice cultivation, along with the increase in levels of soil salinity, the suitable land available for paddies is decreasing in the southwestern coastal area. The survey data revealed that of those households who had access to cultivable land, only 30% of them (ranging from marginal to large farm holders) cultivate rice mostly for their own consumption. Having come to expect salinity as the normal state, farmers have developed practices and strategies to cope with this limiting factor for rice production. These include the following:

Selection of appropriate planting season: Rice is moderately sensitive to salinity (Gregorio et al., 1997). In the wet monsoon season (June–October), when about 80% of the total rainfall occurs, the sensitivity of rice to salinity is reduced because of the dilution of salt in the root zone of the standing crop. During this low-salinity period, transplanted Aman is grown alone in newly accreted land (newly emerged land in the estuaries or in the rivers), locally known as char land, or together with shrimp/freshwater fishes in the medium lowlands. In Chila, this has been the widely used cropping practice in response to seasonal changes in salinity since the emergence of shrimp aquaculture in rice fields.

Use of traditional varieties: Based on land type and crop performance during the planting period, farmers choose traditional varieties with low yields over high-yielding varieties (HYV). Farmers in the study area perceived that local varieties are somewhat resistant to moderate salinity, but that HYV are not. However, HYV Aman rice was reported to have been grown by many farmers in medium-high land to highlands. Because of differences in salinity levels, farmers have been found to select different varieties of rice for medium highlands and medium lowlands.

Experiment with salt-tolerant varieties: With the aim of reviving the dwindling rice production in southern Bangladesh, the Bangladesh Rice Research Institute (BRRI) released BRRI dhan 47 in 2007, a high-yielding modern variety suitable for the Boro season that spans from November to May, when soil salinity is high. BRRI claimed that BRRI dhan 47 was a salt-resistant rice variety that could withstand 12–14 dS/m salinity at seedling stage and 6 dS/m in its entire lifespan, and the expected average yield was 6 tons per hectare (BRRI, 2007). The most commonly used unit for measuring salinity is the deci-Siemens per meter (dS/m), and a salinity level >4 dS/m is considered as saline soil (Haque, 2006). Given the presence of the agriculture extension services offered by both the Department of Agriculture Extension (DAE) and the nongovernmental organizations (NGOs), many respondents grew BRRI dhan 47 on a trial basis, using seed brought in by extension workers with the hope that this would reduce the risk of exposure to salinity in the dry season. Evaluation of the performance of the newly released variety, as mentioned by the research participants in the group discussions, was that the variety did not conform to the claims made by the BRRI. Participants reported that the rice grew normally until the flowering stage, where salinity appeared to affect the later stages, causing white tips, unfilled or poorly filled grain in the panicles, and finally a loss of yield. During the later stage of rice planting in the Boro season, the soil salinity in rice fields may be more than the tolerance limit required to avoid salinity injury at that stage. Some research participants did experiment with Binadhan 7 in their fields (wet-season rice), using free seed from the NGOs, but ended with same results. Binadhan 7, released in 2007 by the Bangladesh Institute of Nuclear Agriculture (BINA), is a short-duration variety tolerant to major pests and diseases but not to salinity, and therefore the results of the experiment were not unexpected. These unsuccessful experiments apparently have decreased trust among farmers in Chila in salt-tolerant modern varieties released by the research institutes. The widely held view is that salt-tolerant, high-yielding modern rice varieties are not suitable for their mixed rice–shrimp cultivation practices.

Land leased beyond the village: Although exact reasons for the salinity intrusion in the southwest coast may be debatable, the fact that salinity has started to encroach on fertile agricultural land is a very real issue for the people. In response to their understanding of the changes associated with the salinity trend, people have been found to exploit the seasonal variability of salinity regimes in their local area; however, people living next to the Pasur River have taken an alternative strategy by gaining access to agricultural land in the Dacope subdistrict, right across the Pasur, where land is suitable for rice and winter crop cultivation. This has been done either through sharecropping or cash-lease arrangements with landlords to gain access to land. Networks of kinship ties or friendship are very important in gaining access to land that the people in this rural area maintain for future reciprocal benefits. By spatially diversifying livelihood activities, it has become possible for households to reduce the potential threats to the households' well-being due to the risks associated with income from agriculture.

7.6 USE OF CLIMATE INFORMATION

In Chila, most of the respondents surveyed (89%) reported using climate information. The daily weather forecast was the sole source of climate information used by the local residents to learn when there was a threat of hazardous weather, which contributes to their knowledge about the risky environment of the area where they operate. Since the study area is frequently affected by tropical cyclones, listening to the weather forecast is considered important by all livelihood groups, especially by the fishing communities for making decisions about fishing in the sea or rivers. The most frequent sources of information were television (42.5%) and radio (41.3%).

The community has adopted the habit of listening to weather forecasts over the years, especially before the start of the tropical cyclone season, and this means that the community is now better informed about tropical cyclone systems. Their alertness created through obtaining climate information, together with the disaster preparedness of the government and the NGOs, has contributed to reducing the cyclone-related death toll to a great extent in the coastal region in recent times. For example, in 2009, almost all of the households in the study area were affected by a severe cyclone, but

Table 7.2 Percentage of households identifying the reasons for not using climate information

Reason[a]	Percentage of household
Do not need it	1%
Do not know where to get it	87%
Do not know how to use it	89%
Do not have access to it	96%
Do not have means to get it	93%

[a]Households cited more than one reason.

there was no loss of human life, as reported by the Chila Union Council (local government organization) during the field research in 2012.

Despite success in reducing mortality from climatic extremes through the use of climate information, there was less evidence of use of climate information for making production decisions to minimize the negative impacts of climate variability. The respondents surveyed cited the reasons being lack of knowledge about the availability of climate information and its use in real-life, and a lack of access and means to obtain climate information, yet they were aware of the importance of climate information (Table 7.2).

7.7 ADAPTATION TO SALINITY ENCROACHMENT IN DRINKING WATER RESOURCES

One of the important changes evidenced in almost every household in the southwest coastal region was the reliance on ponds, rivers, rainwater, and deep tubewell for drinking and other household purposes due to contamination of freshwater sources by saltwater. In Chila, most of the households (99%) use rainwater, along with water from sweet water ponds (Table 5.2). Water may or may not be filtered before drinking. Figure 7.4 shows a concrete tank being used by households for rainwater collection.

Rainwater harvesting is a common practice in the study area. However, only 18% of the surveyed households reported having rainwater-harvesting facilities in which rainwater, collected from roofs or in open areas, is stored in concrete or plastic water tanks. Most of the poor households collect and store rainwater in small earthen pitchers, as they cannot afford to set up the afore-mentioned facilities.

During the drier months, the business of potable water emerges, as many households have to buy drinking water, usually sold in small earthen jars.

Figure 7.4 A rainwater collection tank used by a household.

People reported that they bought jar water from March to May, and for which a three-member household usually spent about BDT90 (US$1.20) per week for 30 L of water. In Mongla Upazila, some NGOs have come into this business recently, with small water desalinization plants where drinking water is supplied in pipes or jars.

7.8 IMPROVEMENT OF SHELTERS: HOUSEHOLDS' RESPONSE TO TIDAL FLOOD

Apart from the various adjustments in agriculture, aquaculture, and livelihood portfolios, households appeared to have invested in improvement of their shelters as the main preventive measure against the perceived increase in tropical cyclones and rising high tides. The survey data revealed that 67% of the households improved or repaired the structures of their dwelling units in the last 5 years to provide better protection against wind, rain, and tidal flood during the event of a tropical cyclone. Households mostly undertook this improvement and/or rehabilitation of their housing facilities after the area had been ravaged consecutively by tropical cyclones in 2007 and 2009. However, making climate-proof housing conditions in this cyclone-prone coastal area entailed substantial investment and technicalities that appeared to be elusive for most households, as demonstrated in the survey data, in which 93% of the dwelling units in Chila were built with only light materials (Figure 4.11).

Another important change recorded in the survey was raising houses to a desired flood protection elevation. From the survey, it appears that about

47% of the households during last 5 years had elevated the dwelling units and/or courtyards to protect their shelters from cyclone-induced tidal floods and inundation during high tides. The elevating technique involved the following: elevating the floor within the house with mud; raising inner and outer courtyards without any changes in the dwelling structure; and raising both the living area and courtyards. People did not follow any particular method or flood management regulation to determine the Flood Protection Elevation (FPE), and were found to have elevated their houses to any height that they thought necessary or of which they were financially capable. However, people who were living in proximity to the river had elevated their houses above the maximum water level (tide water). Participants in group discussions noted that elevating a house was the most economical approach that they could use to prevent tidal floods. Although elevating a house tends to ease the impact of high tides and prevent inundation by high water, it appears to provide little protection against cyclone-induced floods in this coastal area, where a storm tide of approximately 7 m occurs every 5 years (MCSP, 1993).

7.9 MIGRATION

The population census in 2011 revealed that population growth in the coastal zone was markedly below that of the national average growth for the past 10 years (2001–2011) (Table 4.1), contributing to a gradual decrease in population in many coastal districts. For example, Bagerhat district, where this study was conducted, experienced a population reduction of 4.71% in the recent decade (2001–2011), and for Mongla the corresponding figure was 8.35% (BBS, 2011). This demographic record indicates a net out-migration from this hazard-prone area to other parts of the country. Seasonal migration, discussed in Section 4.3.2, provides the means for households to cope with economic stress in slack seasons. From group discussions, it appeared that seasonal migration is more of a survival strategy for poor households than an adaptation to hydro-climatic changes. Aside from seasonal migration, household members make long-term migrations to different locations inside the country area. In Bangladesh, the primary destination of rural migrants is urban centers, accounting for 66%, followed by overseas and rural areas, made up of 24% and 10%, respectively (Afsar, 2003).

There are multiple factors driving migration; however, in Bangladesh, natural hazards are the main cause of migration (Piguet, 2008), particularly flood, drought, and cyclones. These climatic impacts not only result in outright livelihood failure of poor households, but also, in many cases, push people in rural areas to move to urban centers with pull factors including industries and services that offer salaried job and higher household incomes (Herrmann and Svarin, 2009).

Older participants in the group discussion reported that long-term migration from their area to other parts of the country was very infrequent during their youth. However, in the last couple of years, some family members had moved away to different locations of the country, partly because many families were unable to provide sufficient means to survive in this area due to the repeated loss of income from shrimp farming and partly because some people were able to get jobs in formal sectors in cities. These events, along with people's perceptions of increasing cyclonic activities, salinity intrusion, and rising tidal water levels, are likely to have contributed to the impulse for out-migration inside the country.

7.10 DISCUSSION AND CONCLUSION

Rural households are concerned about the changes in the hydro-climatic environment in the southwest coastal region, as their livelihoods, which depend on the natural environment, are being affected (as discussed in Chapter 5). Climate change is perceived mostly as an increase in temperature, a decrease in rainfall, more frequent tropical cyclones, growing salinity intrusion and a rise in tidal water levels (Table 6.1). Although local people perceived a host of changes in the climate variables, climate extremes and their impacts on the physical environment, their responses that represent real forms of adaptation have been primarily focused on the increasing salinity in the environment. Rural households in the southwestern coastal districts have taken steps to respond to these changes by diversifying their livelihoods and changing their farm production practices. Other responses to climate variability and change are more a case of developing coping strategies rather than taking adaptation measures.

The diversification of livelihoods is a common strategy used by households in rural areas of developing countries in anticipation of threats to livelihoods' security due to shocks and stressors (Ellis, 2000). In Chila, households diversified their livelihoods by incorporating aquaculture, agriculture, fishing, raising livestock and poultry, waged employment, business, international remittance, and harvesting forest resources into their portfolio of activities. As expected, diversification strategies that vary along the household-income gradient are highly dependent on the size of the operated land, where ownership of land is positively correlated with a household's income (Figure 5.10). Having less opportunity to access land and other resources, poor households derived a substantial proportion of their income from wage labor as their primary income stream, and that was diversified by involving fishing, petty business, and stocking fish in small homestead ponds. These diversification strategies of poor households are not made merely by choice; rather, they are determined by seasonal factors, as seasonality provides different income opportunities in different seasons (Table 4.5 illustrates the different occupational activities in the different seasons in Chila). For example, members of a poor household catch Hilsa fish between August and October, work as a day-laborers in the "food for work program" from November to February, and migrate to the nearby districts for rice harvesting between April and June. Therefore, the diversification strategy of poor families is more of a survival strategy than of risk management—one that aims to increase sources of income as their primary livelihood activities, are hardly sufficient to meet the basic needs of the households.

The capabilities of the diversification of the well-off households are different from their poor counterparts. In Chila, commercial aquaculture is by far the most important source of cash income for better-off farmers, but it appeared to be vulnerable to climate impacts, diseases, and market failure. However, in good years, these households are able to increase their income base, and that can lead to building larger asset portfolios to diversify their activities in formal wage employment and business, which directly buffers the shock of climatic impacts and other stresses.

The livelihood diversification pursued by rural households is not a strategy unique to areas with climatic disturbances; it can be a decision of rural households to improve future livelihood security by taking advantage of currently available income-earning opportunities, as much of the literature has suggested (Ellis, 2000; Thomas et al., 2007). However, a household's income-improving strategies involving a diverse set of activities plays an important role in reducing vulnerability by way of enhancing a household's capacity to cope with and adapt to climate change, including climate variability and extremes.

The disturbance of low-level salinity within the tidal floodplain in the dry season was initially viewed as an opportunity for innovation and development of a profitable mixed farming system that combined brackish water and freshwater aquaculture and the agricultural production systems. However, with the growing salinity intrusion in surface and ground water, the socio-ecological system has experienced a change. The changing environmental conditions present major challenges for households to continue or to improve agriculture production, and this has led households to progressively change their major livelihood strategies, mostly switching from agriculture to commercial brackish water shrimp aquaculture and business. It appears to be those households' planned adjustments in their livelihood strategies in responses to hydrological disturbances, affecting the socio-ecological system. The adoption of an aquaculture-based economic production system in Chila and elsewhere in the coastal region has made it possible for the households to avert crises in relation to income failure from crop production. However, exposure to other climate risks (such as cyclones, tidal floods, intense summer temperatures) remains the same, only being transferred from agriculture to the aquaculture production system. Moreover, new nonclimate shocks appear to intervene in the farmer's newly adjusted livelihoods to undermine this adaptation. For example, since 1995 shrimp farms have been experiencing an infestation of the white spot syndrome virus (USAID, 2006), causing major losses in this sector. In Bangladesh, when a disease outbreak combines with climatic disturbances, it results in severe and sometimes permanent detrimental impacts on the livelihood systems of the shrimp farmer.

Kofinas and Chapin (2009) argue that when society is confronted with stresses or conditions that change directionally and move beyond past experiences, local knowledge plays a critical role in the way of providing information about local conditions or context that is essential for an effective implementation of new solutions. This is true for the local approach to rice production in changing environmental conditions by utilizing the spatial diversity of the landscape as well as the temporal variability of the climate. In Chila, for example, farmers were reported to grow a locally available, salt-tolerant rice variety during the wet season when salinity levels in the soil are at a minimum, and some of these farmers were also found to take leases on agricultural land beyond their village, with the aim of bringing back farm production. However, a household's adaptive strategies for food security, based on local knowledge, did not appear to be a widespread success in Chila, as articulated in the household survey, with only 8% of the total surveyed households reported having engaged in agriculture. The reason could be the limitation of local knowledge, as Batterbury (2008) noted that abrupt and surprising changes may surpass local skills and social memories.

Faced with increased challenges linked to hydro-climatic change, local people tend to incorporate experimentation and innovation into their economic production system in Chila. Although innovation is generally associated with novel technology or products, here, socio-technological innovation involves experimentation, diffusion, entrepreneurial activities, and market formation of old technologies in a changing socio-ecological context that significantly deviates from the usual-context practices. In this transitional socio-ecological coastal landscape, innovation-driven adaptation is occurring at multiple levels of aquaculture production systems. On the horizontal level, for example, commercial mud crab and mono-sex tilapia farming are being established, along with the prevailing brackish water shrimp aquaculture; and a vertical value chain is also emerging for each type of aquaculture farming that ranges from fry collection to hatcheries, to commission agents, to processing and exporting. Despite the importance of the diversification of the aquaculture sector in the amelioration of livelihood risks, this process is still in its infancy and it has not spread within the coastal belt due to a range of factors, including a lack of access to resources, information, technical assistance, and markets.

It is clear that there are a few modifications being undertaken to ongoing aquaculture farming practices with respect to perceived climate change, including climatic variability and extremes. Some forms of these response measures are the intergradation of traditional knowledge with current farming practices, providing new ways to deal with growing changes in the climatic and environmental parameters. For example, shrimp farmers were observed to raise the height of the earthen bands around the shrimp ponds or to place nets to prevent the fish from escaping during floods. Households' rainwater collection practices are another example of the use of traditional knowledge, in which old practices are being recombined with modern techniques to adapt to disturbances and changes. However, it may also be the case that an inadequacy or absence of public infrastructure (e.g., absence of water supply facility and inadequate coastal polders) dictates the choice of actions to be taken at the individual or household level.

Given that households employ a number of adaptive strategies, livelihood insecurity and poverty remain an intractable problem throughout the coastal region. The depletion of freshwater resources is having an immense impact on the society by decreasing food security, restricting the choice of livelihood strategies, and affecting human health, where these are beyond the means of the local community to solve on a micro level. Drawing on this case study, there are clear winners and losers with regard to adaptive strategies, because better-off households are able to radically transform their livelihood strategies to exploit the opportunities resulting from the changing environment that provide the ingredients to develop robust livelihood systems. It is also clear that many poor households lack the capacity to cope during times of major stress and are unable to renew their livelihoods. These vulnerabilities provide the context in which poor households make their livelihood decisions. In the face of climatic stress, for example households often push their members to migrate for survival reasons when faced with a collapse of their regular sources of income after climatic extreme events. Evidence presented here suggests that in order for adaptations to climate change to be effective in the coastal area of Bangladesh, there need to be specific strategies in response to climate-related risks that are combined with strategies addressing the underlying causes of vulnerability by focusing on increasing livelihood resilience.

REFERENCES

Afroz, T., Alam, S., 2013. Sustainable shrimp farming in Bangladesh: a quest for an Integrated Coastal Zone Management. Ocean Coastal Manage. 71, 275–283.

Ali, A.S.M., 2006. Rice to shrimp: land use/land cover changes and soil degradation in Southwestern Bangladesh. Land Use Policy 23, 421–435.

Afsar, R., 2003. Internal migration and development nexus: the case of Bangladesh. Conference Paper Presented at the Regional Conference on Migration, Development and Pro-Poor Policy Choices in Asia, Dhaka, Bangladesh.

Belton, B., Karim, M., Thilster, S., Jahan, K.M., collis, W., Phillips, M., 2011. Review of Aquaculture & Fish Consumption in Bangladesh. Studies and Reviews 2011-53. The WorldFish Center.

BRRI, 2007. Success Stories: Varietal Development. Bangladesh Rice Research Institute. http://www.brri.gov.bd/success_stories/index.htm (accessed 07.11.13.).

BBS, 2011. Community Report-Bagerhat Zila. Population and Housing Census 2011. Bangladesh Bureau of Statistics, Dhaka, Bangladesh.

Batterbury, S., 2008. Anthropology and global warming: the need for environmental engagement. Aust. J. Anthropol. 19 (1), 470–484.

Deb, A.K., 1998. Fake blue evolution: environmental and socio-economic impacts of shrimp culture in the coastal areas of Bangladesh. Ocean Coastal Manage. 41 (1), 63–88.

Ellis, F., 2000. Rural Livelihoods and Diversity in Developing Countries. Oxford University Press, Oxford, UK.

Gregorio, G.B., Senadhira, D., Mendoza, R.D., 1997. Screening Rice for Salinity Tolerance. International Rice Research Institute, Philippines.

Haque, S.A., 2006. Salinity problems and crop production in coastal regions of Bangladesh. Pak. J. Bot. 38 (5), 1359–1365.

Herrmann, M., Svarin, D., 2009. Environmental Pressure and Rural-Urban Migration: The Case of Bangladesh. MPRA Paper No. 12879. UNCTAD, Geneva.

Kofinas, G.P., Chapin, F.S., 2009. Sustainable livelihoods and human well-being during social-ecological change. In: Chapin, F.S., Kofinas, G.P., Folke, C. (Eds.), Principles of Ecosystem Stewardship. Springer Science + Business Media, pp. 55–75.

MCSP, 1993. Summary Report: Multipurpose Cyclone Shelter Project. Bangladesh University of Engineering and Technology. Bangladesh Institute of Development Studies, Dhaka.

Paavola, J., 2008. Livelihoods, vulnerability and adaptation to climate change: lessons from Morogoro, Tanzania. Environ. Sci. Policy 11, 642–654.

Piguet, E., 2008. Climate Change and Forced Migration. Research Paper No. 153. UNHCR Evaluation and Policy Analysis Unit, Geneva.

Swapan, M.S.H., Gavin, M., 2011. A desert in the delta: participatory assessment of changing livelihoods induced by commercial shrimp farming in Southwest Bangladesh. Ocean Coastal Manage. 54, 45–54.

Shelley, C., Lovatelli, A., 2011. Mud Crab Aquaculture: A Practical Manual. FAO Fisheries and Aquaculture Technical Paper No. 569. FAO Fisheries and Aquaculture Department, Rome.

Thomas, D.S.G., Twyman, C., Osbahr, H., Hewitson, B., 2007. Adaptation to climate change and variability: farmer responses to intra-seasonal precipitation trends in South Africa. Clim. Change 83, 301–322.

USAID, 2006. A Pro-Poor Analysis of the Shrimp Sector in Bangladesh. Greater Access to Trade Expansion (GATE) Project. US Agency for International Development.

Zafar, M., Hasan, M.N., 2006. Marketing and value chain analysis of mud crab (*Scylla* sp.) in the coastal communities in Bangladesh. In: Islam, M.A., Ahmed, K., Akheruzzaman, M. (Eds.), Value Chain Analysis and Market Assessment of Coastal and Marine Aquatic Products of Bangladesh.

CHAPTER 8

Livelihood Adaptation to Climate Change: The Role of Policies and Institutions

Contents

8.1 Introduction	123
8.2 Institutional Interventions in Facilitating Adaptation to Climate Change	124
8.3 Social Safety Nets: Public Responses to Cope with Livelihood Disturbances	125
8.3.1 Supporting Households to Cope with Climate Hazards	127
8.3.1.1 Renewal of Livelihoods in the Postdisaster Situation	*127*
8.3.1.2 Supporting Risk-Spreading Strategies through Employment in Public Works	*127*
8.3.1.3 Building Resilience of the Vulnerable Groups to Sustain Shocks and Stresses	*127*
8.3.2 Limit to SSNs to Benefit Vulnerable Communities	128
8.4 Role of NGOs in Promoting Livelihood Adaptation	129
8.5 National Climate Policy and Livelihood Adaptation at the Local Level	132
8.6 Adapting Development Plans and Sectoral Policies	133
8.7 Discussion	134
8.8 Conclusion	137
References	138

8.1 INTRODUCTION

Having analyzed the household-level aspects of livelihood adaptation in the previous chapter, this chapter aims to explore policies, institutions, and practices to understand their roles in helping rural households' livelihoods adaptation to climate variability and change. Rural communities function within formal and informal institutional settings in dealing with both climatic and nonclimatic issues. Faced with climate-related problems, local people interact with institutions or agencies, such as public organizations and nongovernmental organizations (NGOs), in pursuance of support for securing and sustaining their livelihoods.

From the perspective of the formal institutions, the analysis here focuses on the three approaches to vulnerability reduction that have become prominent in recent years—social protection, disaster risk reduction, and climate change adaptation (Arnall et al., 2010). The next section of this chapter

Experiencing Climate Change in Bangladesh
http://dx.doi.org/10.1016/B978-0-12-803404-0.00008-9
123

provides a description of the social safety net mechanisms as they have evolved to support poor people adversely affected by natural disasters, illustrating their linkage to livelihood renewal in the face of climate disturbances. The third section relates to the role of the NGOs as a formal buffer at the local level to risk reduction and climate change adaptation for the rural community. This is followed by a discussion on the institutions and the policy interventions specific to address climate variability and change. The chapter ends with a discussion of the findings and a conclusion.

8.2 INSTITUTIONAL INTERVENTIONS IN FACILITATING ADAPTATION TO CLIMATE CHANGE

To investigate the role of institutional arrangements in promoting livelihood stability in the face of climatic extreme events, change, and variability, the survey began by asking respondents who they believed should take responsibility for responding to climate change effects. The survey revealed that nearly one-half of the respondents (42%) perceived that the current vulnerabilities to climate hazards would require interventions from the government; 25% identified both NGO and governmental actions; and 33% mentioned both individual and governmental actions. This suggests that local communities are increasingly aware that individual households' actions to secure livelihood functions from hydro–climatic hazards is insufficient, and therefore requires external institutional support to enable better adaptation to climate disturbances. Next, the participants were asked about the types of institutional interventions that would enhance the local adaptive capacity to respond effectively to climate variability and change. The respondents identified that the following measures (Table 8.1) were required to be implemented through the involvement of governmental organizations and NGOs to promote local adaptation to climate change.

The survey data illustrate the communities' concerns over cyclone and salinity risks damaging the local resource base on which they depend for their livelihoods. Interventions to support cyclone and salinity risk management could entail water infrastructure development, which cannot be implemented without support from the national government because of the magnitude of the investment. The development of salt-tolerant crop and fish varieties also requires public investment in research and development.

The local communities perceive that unregulated rapid expansion of shrimp farming by replacing rice production or the restriction of water

Table 8.1 Institutional interventions identified by local residents to be required for buffering impacts of climate change

Intervention theme	Intervention required	Responses (% of n) n = 372
Infrastructural development	Building coastal embankments and their regular maintenance	369 (99.7%)
	Building cyclone shelters	368 (99.5%)
	Developing drinking water supply systems	363 (98.1%)
	Controlling salinity	359 (97%)
Salt-tolerant farming practices	Providing salt-tolerant crop varieties	97 (26.2%)
	Supplying salt-tolerant fish varieties	170 (45.9%)
Regulatory control mechanisms	Controlling expansion of shrimp farming	318 (85.9%)
	Removing obstructions to river-flow	358 (96.8%)
Others	Providing crop insurance and fish insurance facilities	143 (38.6%)
	Strengthening poverty eradication programs	356 (96.2%)

flow through building obstructions in the rivers or canals have been causing detrimental impacts on the coastal ecosystems as well as households' food security. The communities assume that these problems can be resolved through the proper implementation of regulatory measures by the relevant government agencies, as well as interventions by local government representatives.

The survey data show that communities are aware of the fact that their limitations in using coping strategies to respond to hydro-climatic hazards are very much linked to a lack of assets or to household impoverishment. Within this context, community people opt for an integrated adaptation approach that effectively addresses multiple stresses, while providing opportunities to reduce poverty.

8.3 SOCIAL SAFETY NETS: PUBLIC RESPONSES TO COPE WITH LIVELIHOOD DISTURBANCES

In recognizing the poor households' vulnerability to climatic, social, and economic disturbances, public social safety nets (SSNs) have been in operation in Bangladesh for more than 50 years. The SSNs are seen as a means to help the poor to cope with transient crises so that they do not fall into destitution. Under the rubric of social protection, safety net programs in Bangladesh have grown considerably in addressing the risks and vulnerabilities relating to temporary food insecurity resulting from seasonality, disaster and crisis, abject poverty, and the special needs of population groups such as elderly, widowed, and disabled individuals (Rahman and Choudhury, 2012). The salient features of the main types of SSN programs are given below (Table 8.2).

Table 8.2 Overview of key social safety net programs (SSNs)

Category	Main features
Allowances	• Targeting vulnerable groups or persons with special needs • Major programs: Old-Age Allowance, Allowances for the Widows and Deserted and Destitute Women, and Honorarium for Insolvent Freedom Fighters
Food security and disaster assistance	• Mostly food assistance programs focusing on enhancing food security • Helping vulnerable food-insecure people to cope with climatic extremes • Priority programs: Open Market Sales (OMS), Food for Work (FFW), Vulnerable Group Feeding (VGF), Test Relief (TR), and Vulnerable Group Development (VGD)
Employment generation	• Top two programs: Employment Generation Programme for the Ultra Poor and National Service • Employment Generation Programme provides cash payments to male and female workers involved in manual labor, such as the building and maintenance of rural infrastructure • National Service program aims to create employment opportunities for unemployed youth 18–35 years of age
Human development and social empowerment	• Increase the enrollment and reduce dropout of students at the primary school level. Programs include Stipend for Primary Students (SPS) and School Feeding Programs (SFP) • SPS transfers cash to children from the poorer households, and SFP distributes nutrient-fortified biscuits to elementary school students • Female Secondary School Assistance Programme (FSSAP) targets growth in girl enrollment at the secondary school level • Redesigned FSSAP extends up to higher secondary level and includes boys from SSN families as well

8.3.1 Supporting Households to Cope with Climate Hazards

Traditionally in Bangladesh, social safety net programs have been used as a mechanism for disaster risk reduction targeting the poorest and most vulnerable people. However, safety net programs have gone through a process of change over the years with an incremental unfolding of policy agenda (Rahman and Choudhury, 2012). In taking into account the multidimensionality of poverty and vulnerability (Arnall et al., 2010), the Government of Bangladesh has adopted an integrated safety net approach through a wide range of instruments that not only assist the poorest and the most vulnerable people in coping with short-term climate disaster but also promote these people's resilience to climate change.

8.3.1.1 Renewal of Livelihoods in the Postdisaster Situation

Social safety nets include ex-post responses to livelihood disturbances, focusing on enhancing the food security of the most vulnerable sections of society and rebuilding critical livelihood assets. To this end, Vulnerable Group Feeding (VGF) and Vulnerable Group Development (VGD) are the two major programs run by the government in the medium term. These programs target the poorest, especially women-headed households, who tend to be at the highest risk for hunger in the aftermath of climatic extreme events, to enable them to overcome food insecurity and to re-establish income-generation capacity. Under these programs, a food-based transfer is combined with a complementary package of development services including health and nutrition education, literacy training, savings, and support in launching income-earning activities (ILO, 2013).

8.3.1.2 Supporting Risk-Spreading Strategies through Employment in Public Works

During agricultural off-seasons and after climatic disturbances, traditional employment and food stocks become scarce in the rural areas in Bangladesh, exposing poor households to the potential threat of livelihood failure. To ameliorate this livelihood risk, the government uses a combination of safety net instruments such as Test Relief and Food for Work (FFW)/Cash for Work that contribute by employing poor people in public works programs (DDM, 2012; World Bank, 2013).

8.3.1.3 Building Resilience of the Vulnerable Groups to Sustain Shocks and Stresses

A portfolio of safety net programs being implemented from the government budget have a long-term vision of supporting the poor and vulnerable members of society in the way of initiating the process of graduation that helps them to strengthen livelihood assets and reduce their vulnerability to climate-induced hazards. Examples include old age allowances; allowances for widowed, deserted, and destitute women; agriculture rehabilitation; micro-credit; women's self-employment; assistance to small farmers and poultry farms; employment generation programs for the ultra-poor; stipends for primary students; community-based adaptation to climate change through coastal afforestation; and comprehensive disaster management programs.

It is important to note that many of the safety net programs do not have an explicit objective of climate risk management, but it is an outcome of the programs or can be a by-product of the programs. Building graduation platforms through savings, assets, training, human development, new livelihood activities, and better preparation for entry into micro-credit (Rahman and Choudhury, 2012) and safety net interventions address the underlying causes of households' vulnerability to climate disturbances. For example, mandatory or voluntary savings of a portion of transfers of the programs, such as VGD, for example, result in the building of financial capital that can be allocated to cope with natural hazard events as well as to adapt livelihoods to climatic hazards. Access to cultivable land through savings from the program supports ownership of livestock either as a benefit package of the program (Rahman and Choudhury, 2012) that facilitate households' livelihood diversification, as well enhancing risk spreading capacities.

8.3.2 Limit to SSNs to Benefit Vulnerable Communities

Although the government is implementing a long list of SSN programms throughout the country with the objective of reducing poverty and helping poor and vulnerable people to cope with risks, only 33% of the poor are covered by the SSNs in terms of participating in at least one social assistance program (BBS, 2011). The survey data in Chila also show a low level of participation in the SSN programs run by government agencies. Of the adult respondents, 1% were receiving an old age allowance, 0.27% participated in vulnerable group development program, and 1.35% reported having received gratuitous relief (small cash and rice) in 2009 when Cyclone Aila struck the coasts.

Another important issue is that the resources that are provided through SSN programs are very limited. For example, the Vulnerable Group Development (VGD) provides a monthly transfer of 30 kg of grain to each beneficiary for 24 months, which is equivalent to approximately BDT930 (US$12) per month (Planning Commission, 2014). A lack of coordination between government and NGOs running similar SSN programs is also evident from the household survey in Chila. About 4% of the surveyed households were found to have benefited from both government agencies and NGOs, whereas large numbers of poor people remained out of the reach of SSN programs. However, local communities in the group discussions acknowledged the fact that, in this food-deficit area, the seasonal employment programs (Food for Work/Cash for Work and Test Relief) play a crucial role in ameliorating food insecurity after disaster. These programs help poor households in coping with the impacts of disaster, especially withering hunger, as well as in diversifying income sources when casual employment is most scarce at a local level.

8.4 ROLE OF NGOs IN PROMOTING LIVELIHOOD ADAPTATION

In Bangladesh, the Nongovernmental organizations (NGOs), frequently dubbed as micro-finance institutions for being providers of microfinance to the poor, facilitate local adaptations to climate hazards by helping poor build livelihood assets, and thereby reducing overall vulnerability, as well as by financing programs that specifically address vulnerability to hydro-climatic hazards (Agrawala and Carraro, 2010). In the study area, more than 10 types of programs are run by 12 NGOs. Table 8.3 outlines the key activities of some of the NGOs working in Chila, which was prepared based on the information provided by the respective NGOs.

The household survey data in Chila showed that 45% of the sampled households were found to be beneficiaries of at least one NGO program, excluding the beneficiaries of NGO-run public works programs (e.g., road and embankment repair). Table 8.4 looks at the NGO-run programs accessed by the surveyed households. The data illustrate that, from the perspective of coverage of the program, disaster relief and preparedness activities are the main thrust of the NGO programs in this disaster-prone area, followed by income generation and livelihood diversification initiatives. The disaster relief and preparedness programs appear to have a good balance between disaster preparedness and postdisaster relief operations. These help the communities to foster a sense of preparedness for climatic extreme events as well as to deal with the impacts of hazardous events.

Table 8.3 An overview of NGO activities in Chila

Name	Year started operation in Chila	Membership	Activities
CBO	2009	277 Members, from age group 18–45 years	• Micro credit service • Training on tailoring, automobiles, and sewing • In-kind assistance to cyclone-affected poor people
Muslim Aid	2010		• Training on disaster management • Construction, repair, and maintenance of coastal embankments • Digging ponds for rainwater collection • Construction of mounds as cyclone shelters
Rupantar	1998	1313 Members are only women	• Awareness raising about climate change • Encouragement to develop women leadership • Establishment of women's rights
Dhaka Ahsania Mission	2009	450 Members, to be poor, disabled, or widowed	• Implementation of programs on disaster preparedness and management • Assistance for alternative livelihoods
Resource Integration Centre (RIC)	2010	350	• Adult literacy programs
Caritas Bangladesh	1983	300	• Agricultural extension program • In-kind transfers (poultry, crop seed, rickshaws)

Table 8.3 An overview of NGO activities in Chila—cont'd

Name	Year started operation in Chila	Membership	Activities
Samaj Progoti Sangstha	2012	300	• Training on disaster risk reduction • Support for alternative livelihoods (crab hatchery, vegetable gardening) • Small maintenance of public and private infrastructure
Bangladesh Nazarene Mission	2011	180	• Awareness creation on disaster risk reduction
World Vision Bangladesh	1988	1329 Members, of whom 70% are men and 30% women	• Emergency survival packages to cyclone-affected people • Implement programs to increase income, improve nutrition, and provide access to education and immunization • Infrastructure development

Overall, the income generation programs contribute to building the assets of the households and provide a means to cope with hydro-climatic disturbances. Some of the livelihood programs such as brackish-water aquaculture, crab farming, and vegetable gardening in hanging pots are contributing to adaptations to changes in the aquatic environment caused by both non–climatic induced and climate-induced factors. Some NGOs have taken initiatives to support disadvantaged communities to build houses wrecked by previous cyclones or at risk from future disaster. These new houses are made with strong materials that are more resistant to wind and storm surge. As a result, housing programs enhance the poor communities' resilience to the impacts of climate change.

Table 8.4 NGO-run programs participated in by survey households

Category of program	Program items	Beneficiary households (% of n), n = 372
Support for income generation coupled with training	Shrimp farming/crab culture	40 (11%)
	Tree plantation/homestead vegetable gardening	18 (5%)
	Poultry raising	10 (2.7%)
	Cattle rearing	3 (0.8%)
Water, health, and sanitation	Training on health and sanitation	10 (2.7%)
	Transfer/building of water tanks	16 (4%)
Disaster relief and preparedness	Training on cyclone preparedness	60 (16%)
	Transfer of house-building materials (corrugated iron sheet)	15 (4%)
	Food transfer (disaster relief)	6 (1.6%)
	Money transfer (disaster relief)	25 (6.7%)
Housing	Building safer house	35 (9.5%)
Others		6 (1.6%)

8.5 NATIONAL CLIMATE POLICY AND LIVELIHOOD ADAPTATION AT THE LOCAL LEVEL

In Bangladesh, climate adaptation planning has become a priority for the government as the country is one of the most vulnerable countries in the world to hydro-climatic extreme events, such as flooding and cyclone-induced storm surges. According to the decision taken in the seventh session of the Conference of the Parties (COP7) of the United National Framework Convention on Climate Change (UNFCCC) in 2001, the Ministry of Environment and Forests developed the National Adaptation Plan of Action (NAPA) in 2005, in which priority actions were identified to meet the country's urgent and immediate adaptation needs (MOEF, 2005). As a short-term approach to adaptation, the NAPA set out 15 measures, comprising eight interventions and seven facilitating activities, to enhance the institutional capacity of the agencies responsible for climate-sensitive sectors, as well as to promote adaptation in climate-affected areas. In the NAPA priority activities, preference is given to measures for coastal areas for the reduction of exposure to climatic hazards and the promotion of adaptation in coastal livelihoods. However, progress is very slow in implementing the NAPA priority projects; so far, only one project has moved forward from the planning document to the implementation stage, with the support from the Least Developed Countries Fund (Ayers et al., 2014). The Community-Based Adaptation to Climate Change through Coastal

Afforestation is the first-priority project of NAPA to be implemented in five coastal districts, aiming to reduce the exposure of coastal communities to the impacts of hydro-climatic disturbances through afforestation programs (CBACC-CF Project, 2014). Under this project, newly accreted land is being afforested with mangrove plants. This newly created forest will accelerate the accretion process as well as act as a natural barrier against tropical cyclones and storm surges. In addition, this project creates opportunities for income generation locally through fishing, farming, and tree planting.

To move beyond the short-term strategy, the Government of Bangladesh formulated the Bangladesh Climate Change Strategy and Action Plan (BCCSAP) in 2009, based on the NAPA experiences. The main focus of the BCCSAP is to mainstream climate change issues into economic and social development through prioritizing climate change interventions and disaster risk reductions.

To finance the implementation of the BCCSAP, two trust funds, one funded by the government (Bangladesh Climate Change Trust Fund) and the other by multiple donors (Bangladesh Climate Change Resilience Fund), have been established. To date, a total of 270 projects have been approved for financing from the Bangladesh Climate Change Trust Fund (BCCT, 2014). Although not all the projects in the portfolio are solely dedicated to adaptation or vulnerability reduction, such as waste reduction, recycling, and reuse in Dhaka and Chittagong, many of them have a close synergy between sectoral development and adaptation needs. With the objectives of securing the livelihoods of the most vulnerable to climate change, a total of 11 projects funded by the Bangladesh Climate Change Resilience Fund (BCCRF) are in different stages of implementation (BCCRF, 2013).

Aligned with BCCSAP, the government has readjusted the previous coastal embankment program, that focuses on agricultural development, to one that catalyzes agriculture growth under normal conditions as well as protects lives and livelihoods against cyclone-induced storm surges under a plausible climate change scenario. The new policy has been reflected in the Coastal Embankment Improvement Project undertaken with financial support from the World Bank and Climate Investment.

8.6 ADAPTING DEVELOPMENT PLANS AND SECTORAL POLICIES

The Bangladesh government has made attempts to integrate climate change adaptation into general development planning and sectoral planning documents. The Vison 2021 and the National Perspective Plan (2010–2021) that set the country's long-term development target as well as the sixth 5-year plan, a mid-term development plan, emphasizes mainstreaming climate change adaptation across sectors (Ayers et al., 2014). Like environmental concerns, climate change is viewed as a cross-cutting issue (Urwin and Jordan, 2008), and, with this in mind, a host of policies and strategies have incorporated climate change issues, including the Agricultural Policy 2010, the National Water Management Plan 2001, and the National Social Protection Strategy (draft), so they can synergize with adaptation requirements.

These polices provide a supportive context for developing climate-proofing programs and implementing those at a local level, either with their own budget or by external funding. For example, the Department of Agriculture Extension has undertaken project-based initiatives in relation to providing disaster risk reduction and climate change adaptation in agriculture, with aims to foster sustainable livelihoods and food security in drought-prone, waterlogged, flood-prone, and coastal areas. Similarly, agricultural research agencies are contributing to adaptation in agriculture by providing farmers with salt-tolerant and submergence-tolerant rice verities.

8.7 DISCUSSION

People living in the coastal areas of Bangladesh are exposed to severe hydrological and climate-related hazards, including tropical cyclones accompanied by strong winds, storm surges of up to 6–7 m, and intense rainfall and salinity intrusion, and often they are the least able to respond to or cope with the associated impacts, resulting in long-term consequences for their livelihood security and well-being. Due to their limited adaptive capacities, households' strategies for coping with risk, in many cases, appear to be inadequate. The coastal communities' historical experiences with vulnerability to tropical cyclones highlights the limitations of individual households' coping strategies in this region, which often provide the context to be trapped in the cycle of poverty. Within this perspective, SSN programs play a vital role as institutional responses in helping the most vulnerable people in the society to deal with shock and stresses.

Climate-induced extreme events are shown to play a critical role in advancing the development of public safety net policies and options. For example, the flood of 1998 brought the Vulnerable Group Feeding (VGF) program into practice, and since then it has become a prominent component in the safety net portfolio (Rahman and Choudhury, 2012).

The bulk of the safety net programs comprise food-based disaster assistance and public works programs, followed by education stipend programs for young children studying at the elementary to higher secondary levels. The other programs that constitute a relatively large SSN allocation in recent budgets include micro-credit and rural employment, climate change, and sectoral development programs. As a result, in addition to improving the food security of vulnerable households, SSN programs effectively increase a household's risk-spreading capacity by providing wage employment for the household members, which in turn can prevent poor households from onerous coping strategies in the face of climate-induced hazards, such as the selling of productive livelihood assets.

SSN programs addressing vulnerability to climatic extreme events have traditionally been associated with recovery response measures, especially emergency responses to natural disasters. However, Bangladesh has been increasingly focusing on consolidating its ex-post disaster assistance and food-security schemes, and broadening the current safety nets to include programs, albeit on a very small scale, for helping the poorest people to reduce their sources of vulnerability to shocks and stresses. It has been increasingly recognized that the established sectoral development policies and poverty reduction strategies are insufficient in the face of multiple stresses, including climate change and extreme events encountered by the poorest and most socially disadvantaged groups. Hence, a desirable SSN policy approach pursued by the government is to provide assets and livelihood supports to the poorest and most socially excluded individuals so that they can manage climate risks and natural disasters without relying on external assistance.

Despite the undeniable benefits of the SSN programs, their intent to enhance the resilience of poor and vulnerable people to climate variability and extreme events on a greater scale is largely eclipsed by the low levels of coverage and small benefit packages provided by the SSN programs. From a management perspective, the lack of coordination among a very large number of agencies, including multiple government organizations, NGOs, and international organizations, the implementing safety net programs and corruption in government as well as NGO-funded programs are the major obstacles to achieving the intended benefits from this social protection mechanism (Khan and Rahman, 2007).

Although SSN programs have largely dealt with traditional risks, new risks have emerged in the southwestern coastal area and mostly remain unaddressed, such as salinity and climate change. Considering the impacts of climate change in exacerbating poverty, a few programs have been added to the SSN menu, but they remain sketchy to date.

The vulnerability to cyclone-induced disasters and a higher percentage of absolute poverty, compared to the rest of the country, have created an institutional space for proliferation of NGOs in the coastal areas to fill certain needs that are not sufficiently addressed by the formal government institutions. The strong presence of NGOs has created a lot of enthusiasm and confidence among the vulnerable poor community, as these agencies are addressing some of the basic and urgent needs of the community (Taher, 2003). In addition to providing microcredit to the poor, NGOs are active in livelihood protection, disaster preparedness, and relief activities. Some of the programs run by the NGOs have no practical connection to adaptation, such as women's leadership and adult literacy. Many other programs assist poor households in reducing vulnerability to climatic hazards by diversifying livelihoods, and in some cases they facilitate livelihood adaptation, as these programs are structured by taking into account the hydro-climatic risks in the local coastal areas.

However, NGO activities have attracted criticism from the local community. Some local community members argue that while making impoverished individuals and groups self-reliant, some people eventually become over-reliant on NGOs, as they often do not realize their full potential to resolve problems on their own. Some NGO activities are project-bound, where the NGO is recruited through the donor/government project to implement the project at the community level. In such a case, the outputs of the project are seldom to be found in the field after the completion of the project, because neither the project proponent nor the NGO mainstream the project activities into their normal portfolio of activities. Notwithstanding some drawbacks, NGOs endowed with organizational skills and flexibility demonstrate effectiveness and potential in undertaking programs that can facilitate adaptation to climate change.

Over the years, the institutional approach in Bangladesh of addressing vulnerability to climatic disturbances has been centered on policies and measures largely pertaining to SSN programs and disaster risk reduction. In this field, climate change adaptation has begun to emerge in recent years in different government policies, strategies, and planning documents as the government recognizes that climate change poses a major challenge for development. The climate change issue has gained huge momentum, to the point that there has been development of several new policy instruments within a short span of time. Climate dimensions have brought in general planning policy documents as well as some sectoral plans and policies. The formulation of the BCCSAP is viewed as an achievement of the government toward adaptation of development to climate change, and this is considered a reference for justifying investment in the public sectors (Hedger, 2011). A long list of projects is being implemented with the funding from Climate Change Fund dominated by water resources projects, and some projects have no link to adaptation or vulnerability reduction. A project-based approach is presently underway in implementing the BCCSAP, which more often is a one-off approach than an internalization of the adaptation programs in the routine activities of the implementing agencies. However, the stand-alone adaptation policies of the government are in too early a stage of implementation to evaluate how successfully these policies translate on the ground.

Many of the institutional interventions perceived by the local communities as being crucial to reducing vulnerability and promoting livelihood adaptations in the coastal regions have been accepted in many ways in important general development plans, climate-sensitive sectoral plans, social protection strategies, disaster risk management, and climate change policy documents. A number of issues related to shrimp farming, such as the unplanned expansion of shrimp farming or shrimp farming by obstructing river flow, have been identified by local people as impeding local livelihood adaptation to environmental and climatic change. Due to inaction by public agencies, local government, and wider society, these problems can be experienced across the southwestern coast as affecting agro-ecosystem and freshwater resources. This illustrates how the climate change issue is linked to other socio-ecological issues in a complex manner and therefore demands multistakeholders' involvement in formulating climate change adaptation policies integrated with other socio-economic development policies.

8.8 CONCLUSION

In Bangladesh, the issue of climate-resilient livelihoods in the coastal area is being addressed in numerous policies and practices encompassing social safety net programs, disaster risk reduction policies, and climate change adaptation policies combined in developmental plans. These vulnerability-reducing initiatives are underway through safety net and cyclone preparedness and recovery programs that involve government, NGOs, and local communities. They are largely focused on short-term interventions to cope with climate hazards and extreme climate events, although a few programs currently added to the portfolio aim to tackle the underlying vulnerability of the households. However, these policies and programs offer benefits to the poor section of the community only, and the coverage of these programs remains low.

Some planned adaptation interventions aligned with the BCCSAP are in an early stage of implementation in the coastal areas, and these projects are expected to reduce the risk of cyclone disaster as well as to create emergent conditions for local livelihood adaptations to hydro-climatic change. Yet the effective implementation of the BCCSAP is a big challenge in Bangladesh, considering the country's culture of partisan politics, as well as lack of accountability and transparency at all levels (Khan and Rahman, 2007). Moreover, the policies in the national document do not necessarily take shape into actual practice and may not facilitate livelihood adaptation in the way in which it was intended (AIACC, 2006). Nonetheless, the study findings suggest that various institutional interventions helped to promote livelihood adaptation, at least to a limited scale, and can flourish on large scales with appropriate integration and effective implementation of social safety net schemes, disaster risk reduction policies, and climate change adaptation policies.

REFERENCES

Arnall, A., Oswald, K., Davies, M., Mitchell, T., Coirolo, C., 2010. Adaptive Social Protection: Mapping the Evidence and Policy Context in the Agriculture Sector in South Asia. IDS Working Paper 345. Institute of Development Studies at the University of Sussex, UK.

Agrawala, S., Carraro, M., 2010. Assessing the Role of Microfinance in Fostering Adaptation to Climate Change. OECD Environmental Working Paper No. 15. OECD Publishing.

Ayers, J.M., Huq, S., Faisal, A.M., Hussain, S.T., 2014. Mainstreaming climate change adaptation into development: a case study of Bangladesh. WIREs Clim. Change 5, 37–51.

AIACC, 2006. Vulnerability and Adaptation to Climate Variability and Change: The Case of Farmers in Mexico and Argentina. Assessment of Impacts and Adaptations to Climate Change (AIACC), Project No. LA-29. The International START Secretariat, USA.

BBS, 2011. Community Report-Bagerhat Zila. Population and Housing Census 2011. Bangladesh Bureau of Statistics, Dhaka, Bangladesh.

BCCT, 2014. Bangladesh Climate Change Trust. Ministry of Environment and Forests, Government of Bangladesh. http://www.bcct.gov.bd/images/270514/Total%20projects%20fund%20allowcation%20from%202014.pdf (accessed 24.05.14).

BCCRF, 2013. Bangladesh Climate Change Resilience Fund. Government of Bangladesh. http://bccrf-bd.org/Project.html (accessed 24.05.14).

CBACC-CF Project, 2014. Community Based Adaptation to Climate Change through Coastal Afforestation in Bangladesh. Ministry of Environment and Forest, Government of Bangladesh. http://www.cbacc-coastalaffor.org.bd/ (accessed 07.07.14).

DDM, 2012. Social Safety Net Programme: Programme Implemented by the Department of Disaster Management. http://www.ddm.gov.bd/socialsafetynet.php (accessed 18.05.14).

Hedger, M., 2011. Climate Finance in Bangladesh: Lesson for Development Cooperation and Climate Finance at National Level. Institute of Development Studies, UK.

ILO, 2013. Social Security Department: Bangladesh. http://www.ilo.org/dyn/ilossi/ssimain. viewScheme?plang=en&pgeoaid=50&p_scheme_id=1356 (accessed 22.06.14).

Khan, M.R., Rahman, M.A., 2007. Partnership approach to disaster management in Bangladesh: a critical policy assessment. Nat. Hazards 41, 359–378.

MOEF, 2005. Bangladesh Climate Change Strategy and Action Plan 2009. Ministry of Environment and Forests, Government of Bangladesh.

Planning Commission, 2014. National Social Protection Strategy (NSPS) of Bangladesh. Third Draft, General Economic Division. Planning Commission, Government of Bangladesh.

Rahman, H.Z., Choudhury, L.A., 2012. Social Safety Nets in Bangladesh. Ground Realities and Policy Challenges: Process, Coverage, Outcomes and Priorities, vol. 2. Power and Participation Research Centre, UNDP.

Taher, M., 2003. Review of Local Institutional Environment in the Coastal Areas of Bangladesh. PDO-ICZMP Working Paper No. 018. Programme Development Office for Integrated Coastal Zone Management Plan.

Urwin, K., Jordan, A., 2008. Does public policy support or undermine climate change adaptation? Exploring policy interplay across different scales of governance. Global Environ. Change 18, 180–191.

World Bank, 2013. Project Appraisal Document: Safety Net Systems for the Poorest Project. Human Development Unit, South Asia Region, The World Bank.

CHAPTER 9

Conclusion

Contents

9.1 Introduction	141
9.1.1 Summary of the Major Findings	141
9.1.2 Policy Options and Recommendations	146
9.1.2.1 Using a Sustainable Livelihood Approach	*146*
9.1.2.2 Integrating Adaptation and Disaster Risk Management	*146*
9.1.2.3 Safety Nets Should Remain a Key Policy Priority for Vulnerable Households	*147*
9.1.2.4 Public Investment for Society's Adaptations	*147*
9.1.2.5 Major Attention Should Be Paid to Integrated Coastal Management	*148*
Reference	148

9.1 INTRODUCTION

The focus of this study has been to improve the understanding of how people adapt their livelihoods in response to climate variability, change, and extreme events. This research, seeking to empirically examine the process of livelihood adaptation at the household level, has been set in the socio-ecological context of coastal regions of Bangladesh facing multiple shocks and stresses, including those associated with climate variability and change. Drawing on qualitative and quantitative data at the household level, this study allows a better understanding of the complexity of households' adaptation strategies to climate disturbances while providing a conceptual framework that offers generalizations in the analysis of livelihood adaptation in a vulnerable setting of the developing world. In concluding this study, this chapter reviews the major findings of this research and discusses policy implications of the findings.

9.1.1 Summary of the Major Findings

The conceptual framework for livelihood adaptation to climate variability and change set out in Chapter 2 provides the foundation for analyzing the nuances of coping and adaptation strategies that households pursue to respond to climatic hazards. Built on the foundation provided by the sustainable livelihoods approach (SLA), and integrated with the cognitive-based decision-making approach, the framework features basic factors and their relationships within the livelihood systems that shape households' adaptation to the effects of climate change. Livelihood adaptation, in this framework, at the household level involves incremental adjustment or total transformation of livelihood systems to cope with short-term weather-related disturbances and longer-term climate change. The household's access to livelihood assets enables the households to undertake adaptive actions triggered by their perceptions of the climate risks while institutional and policy initiatives play an important role in influencing this process of adaptation. The value of this tool lies in its simplicity, as it comprises a broad set of variables and their linkages, making it easy to apply to specific cases. These are of key importance to understanding climate change adaptation in the complex livelihood-vulnerability context. The study has made this framework operational in the context of rural communities in a southwestern coastal district in Bangladesh.

The investigation of adaptation processes that are part of the livelihood strategies adopted by vulnerable households begins with an understanding of the social, economic, and physical contexts of the coastal area, and then of the livelihood assets of the households as they provide an enabling environment for adaptive actions. In the coastal area in Bangladesh, the biophysical environment mostly dictates the livelihoods of the rural communities, as they are excessively dependent on natural resources for making their living through the direct harvesting of natural resources, such as fishing or the extraction of forest products, and by managing natural resources such as shrimp farming and agriculture. The socio-economic landscape in this area is dominated by the poor, the functionally landless, and small-holding farmers who have insignificant amounts of cultivable land, for which farming practices hardly provide sufficient means for the households' welfare to be secured and sustained. In Chila's case, land distribution is extremely skewed, and so is the distribution of income across communities. With limited opportunities to generate income from farming and formal wage employment, landless and marginal land-holding households turn to small business, fishing, and wage labor. Thus, the present socio-economic situation has come to contribute to a concentration of resources in few large landholders' hands, and that brings about a social vulnerability through limiting access to resource use by the majority of the members of the society, which is often connected to marginalization and persistent poverty (Adger, 1999).

Susceptibility and the exposure to the impacts of climate variability, extreme events, and change have compounded the enduring socio-economic vulnerability of coastal communities. In the low-lying coastal area of

Bangladesh, tropical cyclones are the most destructive climate-induced hazards to the communities, killing large numbers of people and damaging properties, and can make lasting impacts on the livelihood assets of the natural resource–dependent rural households as a consequence. The study reveals that interacting with these extreme weather events, other key stresses, including salinity intrusion, water logging, and riverbank erosion, seriously undermine the social well-being by their impacts on food and water security and the natural resource base on which local people depend for their livelihoods.

This research has sought to analyze the local communities' perception of climate change as a crucial contributor to making adaptive changes. People in the case study believe that the climate has already changed, which is, in their view, manifested in rising temperatures, declining rainfall, increasing cyclone storms, growing salinity intrusion, and rising high tide levels. These assessments of climate change are fairly consistent with the climatological explanations of climate change. The study shows the insights from occurrences of extreme weather, extreme climate events, and hydrological changes in the region appear to form the perception of climate change. The people's attribution of these hydro-climatic events to climate change has been based on their repeated personal experiences and the associated impacts of the events on their lives and livelihoods, which have tended to shape their climate risk perceptions. In the study region, where shrimp aquaculture is the principal livelihood, people are most likely to associate the risks posed by climate change with those climatic phenomena that adversely affect shrimp farming, although they affect other spheres of their lives as well.

Given that the people in Chila perceive a number of changes in the climatic variables, phenomena related to climate and weather, and their impacts on the physical environment, the adaptation measures by these people are mostly linked to the growing salinity in the environment. The evidence that emerged from the study suggests that these adaptations, amid increasing salinity intrusion in soil and water, have been occurring due to the diversification of livelihoods, changing farm production practices, and transforming economic production systems. The diversification process in Chila appears to be dictated by the households' access to cultivable land, the distribution of which is extremely skewed toward the wealthy. Therefore, the diversification strategies undertaken by the poor are largely survival strategies, and consequently are not geared to cope with climate hazards, whereas the deliberate diversification strategies of the better-off farmers directly buffer the shock of climatic impacts and other stresses.

The changing environmental conditions resulting from the growing salinity ingression present major challenges for households to continue with their previous agriculture-based livelihood systems. In response to this environmental change, households have progressively changed their major livelihood strategies principally through switching from agriculture to commercial brackish water aquaculture and related businesses. Ironically, the adoption of aquaculture-based livelihood systems, which contributes to averting income failure from agricultural production caused by salinity intrusion, is one of the causes of growing salinization in the soil and water, and is having an impact on food and water security as a consequence.

Confronted with the changing environmental conditions, people have used their knowledge about the local conditions. Experienced farmers appear to notice the salt-tolerance capacity of locally available rice varieties and the dilution of soil salinity during the wet season, and have used this information for the adaptation of rice cultivation. However, this local knowledge-based adaptive strategy has not emerged as a widespread success, suggesting that the abrupt changes in the environment may have surpassed local skills. It could also be the case that inadequate or absence of public initiatives (e.g., absence of water supply facilities and inadequate coastal polders) limits the choice of actions at individual or household level. The adaptive changes in livelihood strategies are planned adaptation to hydro-climatic disturbances and have clear winners because only a few better-off households are able to exploit these strategies to make resilient livelihoods.

This study has shown that there are components in a number of policies and programs pertaining to social safety nets (SSNs), disaster risk reduction, and climate change adaptation that involve government and nongovernmental organizations. They address the simultaneous reduction of immediate and underlying causes of vulnerability, as well as livelihood adaptation to climate variability, extreme events, and change. There have been institutional initiatives implemented that combine SSNs and disaster risk reduction programs (SSN-DRR programs) that have helped not only to cope temporarily during climate-induced disasters but have also played a role in contributing to local-level adaptation. Although they are on a small scale, these policies provide the poor and marginal households with the livelihood supports for renewal of their livelihood strategies in ways that help them to manage climate risks and natural disasters without counting on external assistance. Recently, SSN policies have adapted further by recognizing the impacts of climate change in undermining existing poverty reduction initiatives; in line with this change, SSN programs have been modified with the addition of climate change adaptation projects that use social protection frameworks to promote livelihood adaptation in the coastal area. Despite the potential of SSN programs to enhance the livelihood resilience of the poor to climate variability and extreme events, the programs' impacts are considerably low because of their low levels of coverage with very small benefits.

In the vulnerable swathe of the coastal region, a large number of NGOs have been active, providing some basic and urgent assistance to the vulnerable communities that are not sufficiently met by the government programs. Targeting poor and socially disadvantaged people, NGOs offer a wide range of services, from covering broader socio-economic issues to exclusively coastal issues. Given the numerous programs run by the NGOs, some of the programs have no link to adaptation, as expected, whereas many other programs contribute to facilitating livelihood adaptations, as the hydro-climatic risks have been considered in the design of these programs. Therefore, managing an effective relationship with NGOs is considered an important social asset, especially by the poor in their pursuit of resilient livelihoods to climatic hazards.

Unlike SSNs, disaster risk reduction and climate change adaptation comprises a relatively new policy framework in Bangladesh. In recent years, the government of Bangladesh has given impetus to the climate change issue in the backdrop of the government's recognition of climate change as a major challenge to poverty reduction and development. In the context of the political signals of the government about the seriousness of climate change, the government's policy instruments have started to adapt to a changing policy context with a two-pronged approach: integrating climate change adaptation into existing sectoral and general planning policies, without changing the scope of the respective policy; and developing a new set of policy instruments on climate change adaptation encompassing both general and sectoral developmental issues, of which the most important is the Bangladesh Climate Change Strategy and Action Plan (BCCSAP). Although it incorporates adaptation elements into existing policy frameworks to support localized adaptation processes, it has made little contribution to this end as yet. Although in the early stages of implementation, some adaptation interventions aligned with BCCSAP have demonstrated the potential to contribute to promoting local livelihood adaptation to climate change in the coastal area, such as the coastal embankment improvement project and community-based coastal afforestation project. The government's mandate for climate change adaptation will continue to make progress on formulating policy instruments and undertaking projects, but the greatest challenge is how well they will succeed in facilitating the local adaptation processes.

9.1.2 Policy Options and Recommendations

As previously noted, this research is motivated by concern for understanding how households in a vulnerable setting cope with, and adapt their livelihoods to, climatic variability, change, and extreme events by focusing on the key components and linkages that exist with vulnerable livelihood systems. A wide range of findings outlined above emerged from an examination of the complexity of the coping and adaptation strategies on a household level. Based on these findings the following recommendations are made for policy makers.

9.1.2.1 Using a Sustainable Livelihood Approach

A sustainable livelihood approach (SLA) framework, integrated with the cognitive element of climatic risk perception, provides a useful framework for conceptualizing the climate change adaptation process at a micro or household level in developing countries. It is apparent from this study that the modified SLA can effectively be utilized for detailed understanding of main components and critical links in livelihood systems that influence adaptation process. This can help policy makers to identify the main causes of climate-induced vulnerabilities in different places, adaptation measures that households have already taken, and factors that facilitate or enable adaptation process within livelihood contexts. The SLA involves participatory processes to determine a household's capabilities and priorities, in relation to adaptation to climate variability and change, which is crucial for developing a sustainable climate change adaptation strategy. Through this process, it creates a supportive environment and the opportunities to encourage individuals to undertake adaptation strategies to climatic disturbances. Different parts of Bangladesh have specific differences in their adaptive capacities and propensities to adapt. However, currently Bangladesh has a generic national adaptation strategy, which is a first big step toward creating an enabling environment for adaptation. However, in order to take this process further and to make it effective, region-specific adaptation strategies should be formulated by taking into account the local needs and priorities under the national policy guidelines.

9.1.2.2 Integrating Adaptation and Disaster Risk Management

Rural households in the coastal area of Bangladesh are experiencing pressures to adapt their livelihoods to environmental changes, including climate change, while facing challenges to deal with weather-related extreme events, especially tropical cyclones. These two issues are connected and present significant challenges to food and water security, poverty alleviation, and sustainable development. Therefore, the responses to these twin challenges, namely disaster risk management and climate change adaptation, would benefit from a more integrated approach within the overall objective of sustainable development. Although this issue is not new, at least at the national level, in practice, project-based discrete measures are being implemented to address climate change under the Ministry of Environment and Forests (MOEF). To mainstream adaptation, which is also a policy of MOEF, the ministry should identify the overlaps and complementariness of the disaster risk management and climate change adaptation approaches, and based on which it can design policy guidelines in a way so that disaster management can contribute to livelihood adaptation in the local socio-economic context.

9.1.2.3 Safety Nets Should Remain a Key Policy Priority for Vulnerable Households

The study suggests that coastal farmers, especially in the southwestern coastal area, have rapidly exited from agriculture and entered into brackish-water aquaculture to adapt their economic production systems to the increasing salinity ingression to the environment. This aquaculture-based livelihood strategy has reduced the diversity of portfolios of income, and therefore any substantial production loss due to climatic or nonclimatic factors traps the poor in persistent poverty. These poor people neither enjoy the benefits from shrimp farming during good seasons, because of their small farm size, nor are able to undertake adaptation measures to a variety of stressors. For this vulnerable group, safety nets can play a crucial role in supporting their production systems through aquaculture and agricultural supply so that vulnerable people do not become the most vulnerable. However, the existing safety net policy agenda focuses on only the most vulnerable households suffering from poverty to smooth consumption or to overcome their transient problems. Therefore, the existing safety net policies should be scaled up by increasing the target groups (both the poorest and the poor) as well as by incorporating the elements of risk management into different economic sectors (e.g., agriculture, fishery, forestry) to ensure inclusive and equitable development.

9.1.2.4 Public Investment for Society's Adaptations

There are multiscale causes of vulnerability in the coastal regions of Bangladesh that warrant interventions at multiple scales. This study illustrates that adaptation measures at a household level are not sufficient to the secure livelihoods and well-being of the coastal societies in the face of hydro-climatic disturbances. Because some of the coastal problems, especially those related to tropical cyclones and salinity intrusion, are orders of magnitude beyond the means of an individual household to solve, and therefore require planned intervention of the national government. Coastal embankments, cyclone shelters, and freshwater supplies are some of these large-scale climate-related actions that can save people's lives and support local livelihood adaptations. However, while implementing large-scale projects, policy makers, practitioners, and members of society should be vigilant about maladaptation, in which projects or programs benefit one group of the society at the expense of increasing the risk to another group.

9.1.2.5 Major Attention Should Be Paid to Integrated Coastal Management

As in many other developing countries, people in the coastal area of Bangladesh are exposed to multiple stresses, including the impacts of climate change. Interacting with the broader set of stressors, climate variability, and change, in a complex way, have impacts on the natural resource base and households' access to livelihood assets, and undermine social well-being. In that case, focusing on adaptation efforts by untangling climate change impacts from the other stressors tends to take attention away from the underlying causes of vulnerability that have impacts on the day-to-day livelihoods of natural resource–dependent communities. In the southwestern coastal region, where salinity front has encroached into the entire landscape, adaptive response at household level via transformation of agriculture into brackish-water aquaculture further contributes to increasing salinization processes. In this case, policy options for promoting adaptation at the micro level are virtually the same as avoiding the main causes of vulnerabilities, while promoting superficial responses. Given this risk of stand-alone adaptation planning, the adaptation approach in the coastal zone should be rooted in an integrated coastal management framework that seeks out win–win situations, whereby actions will meet sectoral needs (e.g., agriculture, fishery, and forestry) and will also enhance the capacity of the coastal communities to respond to the impacts of climatic variability, extreme events, and change.

REFERENCE

Adger, W.E., 1999. Social vulnerability to climate change and extremes in coastal Vietnam. World Dev. 27, 249–269.

GLOSSARY OF TERMS

Adaptation the process of adjustment in natural and human systems to reduce damage or to exploit opportunities to benefit from real or expected climate and associated effects

Adaptive capacity the ability or potential of a system to respond successfully to climate variability and change, including adjustments in both behaviour and in resources and technologies

Coping strategies short-term actions to deal with hazards and livelihood risks

Household a unit comprising those who live in the same compound and also share meals and income with the unit

Institutions structures of social orders that shape political, economic, and social interactions; formal institutions are codified in legally binding documents such as laws, regulations, agreements, and operational arrangements, whereas informal institutions are social or cultural norms and values

Livelihood strategies a range of activates that people undertake to achieve their livelihood goals

Livelihoods the capabilities, assets and activities required for means of living

Maladaptation action taken ostensibly to reduce climate vulnerability of a system that impacts adversely on other systems, sectors, or social groups

Mixed method research the combination of quantitative and qualitative research techniques into a single study

Resilience the capability of a system to self-organize while undergoing change and still retain the same controls on function and structure

Risk a function of probability of hazard events and corresponding losses to be sustained by the exposed people

Sensitivity the degree to which a system is modified or affected by disturbance

Union the lowest administrative unit of government in Bangladesh, consisting of several villages

Vulnerability the degree to which a system is susceptible to, and unable to cope with, adverse effects of climate change, including climate variability and extremes

INDEX

Note: Page numbers followed by "f" indicate figures and "t" indicate tables.

A

Adaptation strategies, 4, 16, 25, 31–32, 86, 106–110, 119–121, 141–142, 146
Adaption, 103
Adaptive capacity, 2, 9–11, 16–18, 20, 24–25, 31–33, 70, 83, 103, 124
Adapt, 2–5, 9, 16, 20, 24–25, 31–32, 34, 66, 70, 103, 118, 120, 128, 141, 145–147
Adjustment, 5, 9–10, 24, 103, 110, 115, 118–119, 141–142
Adverse impacts, 20, 87, 100
Afforestation, 127–128, 132–133, 145
Anthropology, 9–10
Aquaculture, 46–47, 50–51, 58t, 75, 83–85, 97–98, 100, 103–106, 108–111, 115, 117–120, 131, 144, 147–148
Asset, 2–5, 18–21, 23–25, 32, 37, 48, 69–70, 74, 82–84, 86–87, 103, 106, 118, 125, 127–129, 131, 135, 141–143, 145, 148

B

Bangladesh, 1–5, 29–32, 34–36, 42–54, 42f, 49f, 65, 70–71, 73, 75–76, 82, 84, 89, 90f–91f, 91–97, 99, 108–110, 112–113, 116, 118–121, 125–126, 129, 132–133, 133, 133–134, 136–137, 137–138, 141–143, 145–148
Bangladesh Bureau of Statistics (BBS), 42
Bangladesh Climate Change Resilience Fund (BCCRF), 133
Bangladesh Climate Change Strategy and Action Plan (BCCSAP), 133, 136–138, 145
Bangladesh Institute of Nuclear Agriculture (BINA), 112–113

Bangladesh Rice Research Institute (BRRI), 112–113
Bay of Bengal, 43–44, 51–54, 61–62, 94
Behavior, 10, 17, 20–21, 30, 32
Belief, 16, 18, 99–100
BINA. *See* Bangladesh Institute of Nuclear Agriculture (BINA)
Brackish water, 44, 46–47, 59–60, 63, 65, 77–78, 85, 94, 98, 107–110, 118–120, 131, 144, 147–148

C

Capacity, 2–3, 5, 8–11, 13–14, 16, 18, 20, 23–25, 31–33, 66, 70, 83, 103, 108–110, 118, 120–121, 124, 127, 132–133, 135, 144, 148
Capital, 20–21, 24–25, 32–33, 33t, 69–86, 128
CARE, 18–19
Chila, 35–36, 55–57, 55t, 58t, 59–64, 64t, 70–71, 73–75, 77–78, 81–86, 104, 106–114, 107f, 110t, 117–120, 128–129, 130t–131t, 142–143
Climate change adaptation
conceptual framework of, 10–13
adaptive responses, 11–13, 12t
climate stimuli, 10
system definition, 10–11
evolution of approaches, 9–10
key concepts
maladaptation, 15–16
resilience framework in, 14–15
vulnerability, 13–14
livelihood framework, 18–25
assets access, 24
climate-resilient livelihood outcomes, 25
climate risk perception, 24
conceptual framework, 23–25, 23f
embedding climate risk management into, 24

Climate change adaptation (*Continued*)
  feedback mechanisms, 25
  institutional context, 24–25
  sustainable livelihoods approach (SLA),
    19–21, 19f
  process of, 16–18
  motivation, strength of belief as, 16
  socio-cognitive model of, 17–18, 17f
Climate Change Cell, 94–95
Climate data collection, 38–39
Climate hazards, supporting households to
  employment, risk-spreading strategies
    through, 127
  postdisaster situation, livelihoods
    renewal in, 127
  sustain shocks and stresses, vulnerable
    groups to, 127–128
Climate proof housing, 115
Climate stimuli, 10
Climate system, 8, 31–32
Climate threat, 24
Climate variability, 3–5, 9–11, 13–16,
  22–25, 23f, 29–34, 38, 42, 66, 70,
  87–88, 97, 103–104, 114, 117–118,
  123–124, 135, 141–146, 148
Climate variables, 53, 113–114, 117
Climatic change, 3
Climatic factors, 97, 147
Climatic phenomenon, 93t, 100–101, 143
Climatic risk, 1, 24, 108–109, 111, 136,
  145–146
Climatic stimuli, 9–12
Coastal area, 1, 3–5, 33, 34–35, 42–45, 47,
  50–51, 50f, 53, 65–66, 69–71, 75,
  82–83, 85, 89–94, 97, 111,
  115–116, 120–121, 132–138,
  142–148
Coastal communities, 1
Coastal embankments, 98–99, 125t,
  130t–131t, 133, 145, 147–148
Coastal region, 1, 29–30, 34–35, 42f,
  43–44, 51–54, 62, 65, 73–74, 86,
  89, 90f–91f, 99–101, 103, 108–110,
  113–114, 117–121, 137, 141, 145,
  147–148
Cognitive barriers, 18
Cognitive, 4, 17–18, 22–24, 141–142, 146
Commercial aquaculture, 108–109, 118

Community-based Adaptation to Climate
  Change, 127–128, 132–133
Cope, 3, 13–14, 18–20, 70, 83, 111, 116,
  118, 120–121, 125–129, 131, 134,
  137–138, 141–146
Coping strategies, 4, 12–13, 31–32, 83,
  110–111, 117, 125, 127–129,
  134–135, 141–142, 146
Correlations, 94–95, 99–101, 106
Corrugated iron (CI) sheets, 73
Crab fattening, 45, 57
Cultivable land, 61, 75, 84, 108, 111, 128,
  142–143
Cyclone Aila, 63–64, 64t, 128
Cyclones, 1–2, 8, 22, 33t, 34–35, 37–39,
  51–53, 53t, 56, 63–64, 64t, 66, 75,
  83–84, 86, 89–94, 90f, 97–100, 104,
  106, 113–119, 124, 125t, 128,
  130t–132t, 131–134, 136–138,
  142–143, 146–148
Cyclonic disturbances, 89
Cyclonic storms, 51, 52f, 89–94, 90f, 92f,
  93t, 97, 100–101

D
Damage, 9–10, 13, 24, 42, 50–51, 53, 59–60,
  63–64, 64t, 66, 91–92, 98, 98t
Darwin, 9
Data-gathering, 37, 107–108
Decadal frequencies, 89
Demographic indicators, 44t, 55t
Department for International Development
  (DFID), 18–19
Department of Agriculture Extension
  (DAE), 112–113
Department of Fisheries (DOF), 47
Depletion of freshwater resources,
  120–121
Detrimental impacts, 109–110, 118–119,
  124–125
Developing countries, 8, 14, 22, 104,
  117–118, 146, 148
Development plans, 133–134, 137
DFID. *See* Department for International
  Development (DFID)
Disaster, 5, 13–14, 38, 72–73, 82–83,
  113–114, 123–129, 126t, 130t–132t,
  131, 133, 135–138, 144–147

Disturbance, 1, 4, 8, 10–11, 14–15, 20, 22, 31–34, 87–89, 91–92, 97, 103–122, 125–129, 131–133, 136–137
Drainage congestions, 50–51, 53
Drinking water, 73–74, 84, 86, 114–115, 125t

E

Entrepreneurial activities, 119–120
Environmental and socio-economic stresses, 31
Environment, 1–2, 8–9, 11, 19–20, 23–24, 30–31, 34–35, 37, 48, 60, 76–77, 87–88, 93–95, 100–101, 106–107, 110–111, 113, 117, 120–121, 131–134, 137, 142–144, 146–147
Erosion and accretion, 43–44, 53, 65, 142–143
Evaporation rate, 98
Experimentation, 109–110, 119–120
Extreme event, 2, 5, 10, 22, 33–34, 38, 51–52, 70, 87, 92–93, 111, 120–121, 124, 127, 129, 132–135, 141–147

F

Facilitating adaptation, 5, 124–125
Farakkah Barrage, 50–51
Field data collection
  climate data collection, 38–39
  focus groups, 37–38
    hazard mapping, 37–38
    seasonal calendar, 38
  household survey, 37
  introductory community visit, 36
  key informant interview, 38
  secondary information review, 36
Financial capital, 85
  income, 78–80, 78f–80f
  loan, 80–81, 81t
Flood Protection Elevation (FPE), 115–116
Focus groups, 37–38
Fourth Assessment Report of the Intergovernmental Panel on Climate Change (IPCC AR4), 3
Freshwater systems, 43–44, 53, 86, 94, 99, 109–111, 114, 118–121
Future climate change, 31–32

G

Ganges-Brahmaputra-Meghna (GBM), 43–44
Global environmental change, 13–14
Global warming, 8, 94
Greenhouse gas, 8, 15
Group discussion, 37, 62, 93, 97, 111–113, 115–117, 128–129

H

Hazard mapping, 37–38
Hazardous weather, 100, 113
Hazards, 1, 4–5, 10, 13–15, 22, 24, 37–38, 42, 50–52, 66, 82, 100, 106–111, 113, 116–117, 124–125, 127–129, 132–138, 141–143, 145
High-yielding varieties (HYV), 112
Hilsa, 61
House hold human and natural system, 11–12, 15
Human capital, 83
  education, 71, 71t
  health, 72, 72f
  household size, 70–71, 70f
  technical training, 72–73, 72f
Human casualty, 92
Human society, 4, 24, 70
Human systems, 9–12
Hydro-climatic exposure, current vulnerability to
  land erosion, 65
  tropical cyclone, 63–64
  water and soil salinity, 62–63, 63f
Hydro-climatic hazards, 1, 4, 42, 66, 124–125, 129
Hydro-climatic stressors, 108–109, 116, 125
Hydro-climatic variability, 88–93
  temperature/precipitation variability, 88, 88f–89f
  tropical cyclones, 89–93, 90f–91f
Hydro-climatic variations, 3, 87–93, 100
Hydro-meteorological records, 87–88
Hydro-meteorological trends, 94
HYV. *See* High-yielding varieties (HYV)

I

Income distribution, 78–80, 78f–80f
Income-earning opportunities, 106, 118

Incremental, 5, 24, 103, 127, 141–142
Informant interview, 37–38
Innovation, 109–110, 118–120
Institutional responses, 134
Institutions
  formal, 21
  informal, 21
Intergovernmental Panel on Climate
    Change (IPCC), 3, 9–10, 13–14
Inundation, 50, 53, 64, 91–92, 94, 97, 110,
    115–116
IPCC. *See* Intergovernmental Panel on
    Climate Change (IPCC)

L

Land erosion, 65
Land holding size, 46f, 50–51, 53t
Landuse, 37
Livelihood activities, 4–5, 24, 32, 42–43, 45,
    57, 58t, 61, 82, 100, 103–104,
    106–107, 107f, 113, 117–118, 128
Livelihood adaptation, 2, 4–5, 8–9, 23–25,
    23f, 29–30, 32–34, 36, 103,
    141–142, 144–148
  diversification for, 104–106, 104f–105f
  institutional interventions, 124–125, 125t
  migration, 116–117
  national climate policy, 132–133
  rice production, salinity intrusion in,
      111–113
  shelters improvement, 115–116
  shrimp aquaculture, coping strategies in,
      110–111, 110t
  social safety nets, 125–129, 126t
    climate hazards, supporting households
        to. *See* Climate hazards, supporting
        households to
    development plans and sectoral
        policies, 133–134
    NGOs in, 129–131, 132t
    programs addressing vulnerability, 135
    vulnerable communities, 128–129
  strategies, 106–110, 107f
    aquaculture, 107–108
    commercial aquaculture, traditional
        practices incorporation into, 108–109
    risk-spreading, new species as,
        109–110

Livelihood assets, 3, 5, 20, 23–25, 37, 86–87,
    106, 127–129, 135, 141–143, 148
Livelihood capitals
  financial capital. *See* Financial capital
  human capital. *See* Human capital
  natural capital. *See* Natural capital
  physical capital. *See* Physical capital
  social capital. *See* Social capital
Livelihood decisions, 24, 120–121
Livelihood diversification, 103–106,
    104f–105f, 118, 128–129
Livelihood group, 36–37, 46, 113
Livelihood portfolios, 108, 115
Livelihoods, 1–5, 8, 13–14, 16, 18–25, 19f,
    29–30, 32–38, 33t, 42, 45–46, 48,
    50–51, 57, 58t, 60–62, 65–66,
    70–82, 87, 97, 98t, 100, 103–110,
    115, 117–118, 120–121, 123,
    125–133, 136, 138, 141–148
Livelihood strategies, 2, 4–5, 19–25, 37, 59,
    62, 76, 80, 86, 103–110, 118–121,
    142, 144, 147
Livestock farming, 76–78, 77f
Local communities, 3, 32–34, 36, 85–87,
    100–101, 120–121, 124–125,
    128–129, 136–138, 143
Local perceptions, 37–38, 87–88, 93–94,
    93t, 98t, 99, 106–107
Long-term climate change, 24, 38

M

Maladaptation, 2, 8–9, 15–16, 18,
    147–148
Malnutrition, 45
Meteorological records, 94–95, 99–101
Migration, 38, 44–45, 60–62, 116–117
Ministry of Environment and Forests
    (MOEF), 146–147
Mixed methodologies, 31
Mixed method research, 29–31
Model of Private Proactive Adaptation to
    Climate Change (MPPACC),
    17–18, 17f, 24
Mongla, 35, 38, 53–55, 54f, 57, 62–63, 63f,
    88f–89f, 88, 95–97, 99, 114–116
Monoculture system, 108–109
Monsoon, 48–51, 53, 63, 73–74, 88, 93t,
    94–98, 96f, 111

Most vulnerable, 1, 15, 31, 53, 83, 127, 132–134, 147
Moving average, 95
Mud crabs, 108–110, 119–120
Multiple stresses, 125, 135, 148
Multiple stressors, 86

N
NAPA. *See* National Adaptation Plan of Action (NAPA)
National Adaptation Plan of Action (NAPA), 132–133
National Climate Policy, 132–133
National Perspective Plan, 133–134
Natural capital, 84–85
   land, 75–76, 75f–76f
   livestock, 76–78, 77f
Natural systems, 11–12, 15
Non-farm activities, 106
Non-governmental organizations (NGOs), 2, 33t, 34–35, 48, 71–73, 80–82, 85–86, 112–115, 123–124, 129–131, 135–138, 145

O
On the Origin of Species (Darwin), 9

P
Perceptions, 3–5, 17–18, 23–25, 32, 37–38, 103, 106–107, 117, 141–143, 146
   climate change, 93–94, 93t
   meteorological information, 94–97, 96f
   risks, 97–99, 98t
   hydro-climatic variability. *See* Hydro-climatic variability
Physical capital, 83–84
   cyclone shelters, 75
   housing, 73, 73f
   production equipment, 74, 74t
   safe water, access to, 73–74, 74t
Planting season, 111
Policy, 2–3, 5, 11–12, 14, 18–19, 22, 34, 36, 123–124, 127, 132–135, 137
   options and recommendations
      integrated coastal management, 148
      integrating adaptation and disaster risk management, 146–147

society's adaptations, public investment for, 147–148
   sustainable livelihood approach (SLA), 146
   vulnerable households, safety nets for, 147
Policy response, 2
Polyculture system, 108–109
Population, 4, 13, 35–36, 44–45, 48, 50–51, 53–56, 55t, 56f, 59–60, 65, 70–71, 73, 74t, 78–79, 84, 99–100, 116, 125–126
Poshur River, 53–55
Poverty, 2, 8, 16, 18–19, 22, 45, 47, 65, 71, 85–86, 111, 120–121, 125–128, 134–136, 142, 144–147
Prolonged exposure, 98
Protection Motivation Theory (PMT), 17

Q
Qualitative and quantitative research, 29–31
Questionnaire, 37

R
Rainfall, 16, 18, 22, 48–50, 62–63, 73–74, 88, 89f, 93t, 94–98, 96f, 100–101, 106, 110–111, 117, 134, 143
Rainfall regime, 88
Rainwater collection tank, 115f
Rainwater harvesting, 114, 115f
Rehabilitation, 82, 115, 127–128
Research method
   case studies, 34–36
      communities, selection of, 35
      sample households, selection of, 35–36
      scoping of, 34–35
      selection, 35
   conceptual and technical design
      climate variability and extremes, 31–32
      institutional analysis framework, 33–34
      sustainable livelihoods approach, 32–33, 33t

Research setting
  Bangladesh, coastal region of
    climate change, potential
      impacts of, 53
    climate, characteristics of, 48–50, 49f
    physical setting, 42f, 43–44, 44t
    salinity intrusion, 51
    socio-economic context. *See*
      Socio-economic context
    tropical cyclone and storm surge,
      51–52, 52f
    waterlogging and drainage
      congestions, 50–51, 50f
  study area, description of, 53–65, 54f
    agriculture, 61
    farming, 57–60, 59f
    fisheries, 60–61
    geography of, 55
    health and education facilitation,
      56–57, 56t
    hydro-climatic exposure, current
      vulnerability to, 62–65
    livelihood activities, 57, 58t
    population dynamics, 55–56,
      55t, 56f
    public services, 57
    road infrastructure, 57
    seasonal migration, 62
    Sundarbans forest resources,
      extraction of, 61–62
Resilience framework, 15
Resilience, 2, 5, 11, 14–15,
    22–23, 29–33, 120–121,
    127–128, 135, 144–145
Response, 2, 9–13, 15–16, 18, 22, 24–25,
    32–34, 37, 103–104, 113, 118–121,
    125–129, 125t, 134–135, 141, 144,
    146–148
Rice production, 111–113, 119, 124–125
Risk appraisal, 17–18
Risk management, 4, 13–14, 23–25,
    117–118, 124, 128, 137, 146–147
Risk reduction, 33–34, 72–73, 82,
    123–124, 127, 130t–131t,
    133–134, 136–138, 144–145
Robustness, 120–121
Routine, 5, 12t, 24, 103, 136–137
Rural livelihood, 2

S
Safety net programmes, 85–86, 125–128,
    126t, 135, 137–138
Salinity intrusion, 51, 66, 76, 94, 97,
    99–101, 111–113, 117–119, 134,
    142–144, 147–148
Saltwater intrusion, 73–74, 84, 94, 99
Samity, 81–82
Sample size, 30, 35–36
Scales, 4, 8, 11–13, 15, 22, 24–25, 53, 60, 77,
    95, 135, 138, 144–145, 147–148
Scientific evidences, 94, 99–101
Seasonal calendar, 38
Seasonal migration, 62, 116
Sectoral policies, 133–134
Sensitivity, 10, 13–14, 111
Short-term variability, 38, 100
Shrimp aquaculture, 46–47, 57, 59–61,
    64–65, 74, 76, 84, 98–100, 105–108,
    110–111, 143
Shrimp farming, 37, 44–47, 47f, 55, 59–60,
    74, 78, 80, 94, 97, 99–100, 104–106,
    108–111, 110t, 117, 124–125, 125t,
    132t, 137, 142–143, 147
Shrimp ponds, 46–47, 57, 60, 64, 64t,
    98–99, 109–111, 120
Shrimp value chain, 111
Sidr, 92–93
Siltation, 50–51
SLA. *See* Sustainable livelihoods
    approach (SLA)
Social capital, 85–86
  community organization, participation
    in, 81–82
  NGOs, contact with, 82
Social-ecological systems (SES), 2, 11,
    14–15
Social safety nets (SSNs), 125–129, 126t,
    144–145
Society-nature interface, 31
Socio-cognitive, 4, 17–18, 24
Socio-economic context
  fishing, 47
  institutional environment, 48
  land distribution, agriculture and
    structure of, 45–46
  livelihood characteristics, 45
  NGOs, role of, 48

population, 44–45, 44t
poverty and unemployment, 45
shrimp farming, 46–47, 47f
Socio-political-economic system, 31–32
Soil salinity, 51, 62–63, 77–78, 85, 97,
    111–113, 144
Southwest coast, 1, 35, 47, 73–75, 85,
    108–109, 113–114, 117
SSNs. *See* Social safety nets (SSNs)
Storm surges, 1, 44, 50–53, 53t, 63, 65–66,
    75, 83, 91–92, 93t, 94, 97–99,
    132–134
Storm tide, 91–92, 115–116
Study site, 42–43
Subjective judgement, 87
Summer temperature, 49–50, 88, 98t,
    118–119
Sundarbans, 43–45, 53–55, 57, 61–62
Surge heights, 53t, 92–93, 92f, 97
Survey sustainability, 86
Sustainable livelihoods approach (SLA), 4,
    19–21, 19f
    assets, 20
    institution and structures, 20–21
    strategies and outcomes, 21
    vulnerability context, 20
Sustainable livelihoods, 19–21, 19f,
    32–33, 48

T
Tenualosa ilisha. See Hilsa
Third Assessment Report (TAR), 10, 13–14
Threshold, 8, 18
Traditional varieties, 112
Transformations, 21, 24, 103, 141–142, 148
Trend analysis, 97

Trigger, 5, 11–12, 16, 18, 24, 32, 103, 141–142
Tropical cyclone, 1, 8, 34–35, 53, 63–64, 66,
    75, 83, 86, 89–93, 93t, 97–99, 106,
    113, 115, 117, 132–133, 134,
    142–143, 146–148
    and storm surge, 51–52

U
Unemployment, 45
Unequivocal, 100–101
Union Welfare Center (UWC), 56–57
Unique, 1, 18–19, 45, 65, 118
United National Framework Convention
    on Climate Change (UNFCCC),
    14, 132–133
UWC. *See* Union Welfare Center (UWC)

V
Vison 2021, 133–134
Vulnerability, 1–2, 5, 11, 13–14, 14f, 19–20,
    24–25, 31–32, 35, 42, 61–65, 69–70,
    86, 98–99, 103–104, 108–109, 118,
    123–126, 128–129, 136–138,
    142–145, 147–148
Vulnerable Group Development (VGD),
    126t, 127–129
Vulnerable Group Feeding (VGF), 127

W
Wage employment, 80, 85, 104–105, 118,
    135, 142
Wage labourers, 57, 80, 105–106,
    117–118, 142
Warmer, 93t, 94–95
Waterlogging, 50–51, 50f, 53
Winter, 49–51, 61, 88, 94–96, 96f, 99, 113

Printed in the United States
By Bookmasters